PROSPECTS FOR SCHOOL MATHEMATICS

edited by

Iris M. Carl
Mathematics Education Trust

MATHEMATICS EDUCATION TRUST
established by the
NATIONAL COUNCIL OF TEACHERS OF MATHEMATICS

Library of Congress Cataloging-in-Publication Data:

Seventy-five years of progress : prospects for school mathematics /
   edited by Iris M. Carl.
      p.   cm.
   Includes bibliographical references.
   ISBN 0–87353–418–2
      1. Mathematics—Study and teaching.   I. Carl, Iris M., date .
   QA12.S465   1995
   510′ .71—dc20   95–10255
                                                              CIP

Financial support for the development of this publication has been provided by contributions to the Mathematics Education Trust, a foundation established in 1976 by the National Council of Teachers of Mathematics. It provides funds for special projects that enhance the teaching and learning of mathematics.

Partial support for the printing and distribution of this publication has been contributed by the Exxon Education Foundation in honor of the NCTM's seventy-fifth anniversary and as its partner in efforts to advance the reform of mathematics education.

Printed in the United States of America

Table of Contents

PREFACE . . .

THE forces changing today's world have had a tremendous impact on the quality of education in our society. Schools that once routinely produced students adequately prepared for productive citizenship are today engaged in making substantive changes in their curricula, assessment, programs, and structure to regain that viability. With inconclusive evidence of progress, the educational reform movement that has dominated the national agenda for more than a decade is intensifying and becoming more urgent.

An early response to the call for reform from the mathematics community was the consensus document *Curriculum and Evaluation Standards for School Mathematics*, developed by the National Council of Teachers of Mathematics (NCTM). This coherent vision performed an indispensable function—to significantly improve the teaching and learning, and their associated evaluation, of mathematics—and will be enhanced by its companion publications, *Professional Standards for Teaching Mathematics* and *Assessment Standards for School Mathematics*.

Seventy-five Years of Progress: Prospects for School Mathematics offers a broad view for a varied audience of what the vision expressed in all three *Standards* documents will look like when aligned with practice and outlines ways in which desired changes can be realized. It is a focused work approached from a variety of perspectives on what students and their teachers should know and be able to do. No specific background or knowledge is needed to understand the contents of this book. Through its language, style, and tone it informs the newcomer, skeptic, and advocate and invites curious, demanding readers to consider the many sides of pervasive issues affecting the attainment of the goals for excellence in education. By design it guides the architects of change who seek to afford every student genuine access with equity to a quality education.

Twenty-one essays are organized in four major sections—Students, Teachers and Teaching, Content, and Context—with each section discussing specific topics that explore prospects for advancement into the twenty-first century. These topics were specially prepared for this volume by a select group of distinguished educators, researchers, and policy analysts in education, business, and government. These essays examine the enlightened as well as the misconceived attitudes and ideologies about teaching, learning, schooling, and mathematics as a discipline while conveying and responding to the major concerns of change agents, decision makers, educators, parents, and students. Each paper attests that standards for excellence and equity are not incompatible.

The connection among the sections is the collective focus on students and the fusion of shared perspectives. In every chapter students are both the subject and the object of the reform debate. They animate the data through an extraordinary range of descriptions and personal accounts related by the authors. In the vignettes, students' voices refute beliefs about limited ability, challenge public assumptions, and offer an appreciation for the richness in their diversity and multiculturalism. The prominence of students in the text presents compelling evidence of both their plight and their promise in this advancing technological society.

The common themes emphasized in this volume are (1) teachers' knowledge of content, pedagogy, students, and learning, (2) students' knowing, understanding, and "doing" mathematics and communicating mathematically, and (3) societal and cultural components that affect schools and schooling. Combined with a careful analysis of a K–16 learning-teaching-content-evaluation linkage, much-needed light is shed on the intricacies of reform processes in the pluralistic world of education where schools, the major providers of knowledge and skills, are not the sole providers of learning experiences. A commitment to helping all students become mathematically literate, lifelong learners capable of earning a living and making informed decisions resonates in the closing challenges and recommendations.

The usefulness and applicability of this resource derives from the thoughts expressed by this cross section of experts that yield a wellspring of ideas and new paradigms from different viewpoints. Implicit in their articles is the fundamental premise of the *Standards* documents: that they are not intended to be prescriptive and that state and local determinations, options, and initiatives based on the *Standards'* criteria will be dissimilar. The bibliographies contain provocative sources for further distillation of the insights the authors provide, with the perusal of each entry adding a subsequent supply of references. The cumulative impact of *Seventy-five Years of Progress: Prospects for School Mathematics* is clear images of the transformation envisioned by the three *Standards* documents and the action it will stimulate toward making the NCTM vision a reality.

This book is a commemorative gift to the NCTM in celebration of its seventy-fifth anniversary from the Mathematics Education Trust (MET), the education foundation of the Council, and a product of the ingenuity of the members of the committee who administer the trust funds. I thank them for their financial commitment and moral support.

For many years, the Exxon Education Foundation has been a strong partner in NCTM's efforts to advance the reform of mathematics education. By coincidence, the foundation is celebrating its fortieth anniversary in 1995. As part of the celebrations of both anniversaries, the Exxon Education Foundation has provided partial support for the printing and distribution of this publication. On behalf of the Mathematics Education Trust, I thank Exxon for its generous support.

Many individuals assisted directly in making this resource book possible. Special thanks go to the members of the MET Editorial Panel for their thoughtful development of the project, careful selection of the authors, diligent review of the manuscripts, and thorough critique of the volume. Without their unstinting effort this task could not have been accomplished. Moreover, I gratefully acknowledge the dedication of the authors of this monograph for their contribution of time, expertise, and scholarship that created and shaped the content. The book you hold has been realized because of the invaluable technical assistance of the NCTM Headquarters Office publications staff. To them I extend my sincere gratitude.

—Iris M. Carl

HISTORY . . .

Paths to the Present

Jeremy Kilpatrick & George M. A. Stanic

> *The number of times the teaching of mathematics has been*
> *reformed and the general similarity of view of the reformers are*
> *always interesting to the student of the history of the subject.*
> *(Smith 1922, p. 297)*

A s the first two decades of this century drew to a close, mathematics was increasingly threatened as a school subject. In the elementary school, arithmetic had already come under pressure to reduce the time spent on it, drop unnecessary topics, and shift the emphasis to problems having social utility (Kilpatrick 1992, pp. 13–20). Now the high school mathematics curriculum was in jeopardy. States were steadily decreasing their requirements in algebra and geometry, and enrollments were dropping (Stanic 1986, pp. 195–6).

At the winter meeting of the National Education Association (NEA) in Chicago in February 1919, speaker after speaker stepped to the podium to deliver sharp criticisms of mathematics (Austin 1928). The mathematics teachers at the meeting had no good way to respond. They had no place on the NEA program and no comparable association to serve their interests. Teachers of English had established a national council in 1911, but mathematics teachers had organized only local associations such as the Association of Teachers of Mathematics of the Middle States and Maryland, the New England Association of Mathematics Teachers, and the Central Association of Science and Mathematics Teachers.

When the Chicago Mathematics Club met the following month, the members appointed a committee to correspond with teachers around the country

· · · · · · · · · · · · · · · ·

Jeremy Kilpatrick is Regents Professor of Mathematics Education at the University of Georgia. He has written on the history of curriculum development and of research in mathematics education.

George Stanic is an associate professor of elementary education at the University of Georgia. His historical research concentrates on the U.S. mathematics curriculum.

to learn their views on forming a national organization. About 100 secondary mathematics teachers were contacted, and a large majority of the replies favored the idea. At the next meeting of the NEA, in Cleveland in 1920, 127 mathematics teachers from twenty states met to organize the National Council of Teachers of Mathematics (NCTM) (Austin 1928; Osborne and Crosswhite 1970, pp. 195–6).

A CHANGING ROLE IN REFORM

The NCTM was born out of adversity. Its first president, C. M. Austin (1921, pp. 1–2), acknowledged as much in his report on the establishment of the Council:

> Mathematics courses have been assailed on every hand. So-called educational reformers have tinkered with the courses, and they, not knowing the subject and its values, in many cases have thrown out mathematics altogether or made it entirely elective.... To help remedy the existing situation the National Council of Teachers of Mathematics was organized.

Although the NCTM steadily grew in membership, acquired a journal (*Mathematics Teacher*), and began to issue recommendations for improving the teaching of mathematics, its influence at the national level remained marginal. High school course enrollments continued to decline during the 1930s, and general educational reforms associated with social efficiency, behaviorist theories of learning, and progressive education continued to take their toll on school mathematics into the 1950s (Osborne and Crosswhite 1970, pp. 213–18; Stanic 1986).

The advent of the so-called new math reforms of the 1950s and 1960s put school mathematics on the national agenda and encouraged those who wanted better mathematics instruction and a more prominent place for mathematics in the curriculum. The reforms, however, were not begun by an organization of mathematics teachers. University mathematicians played the dominant role. The NCTM followed the movement, trying to keep its members informed about developments by inviting the reformers to speak at its meetings and to publish in its journals, pamphlets, and yearbooks.

The immediate stimulus for the new math reforms was the concern among mathematicians, scientists, and policymakers that the nation faced a serious shortage of mathematically trained personnel. After some reform efforts were already under way, the Soviet Union launched the first sputnik in 1957. The U.S. government responded by dramatically increasing its support for curriculum development projects and teacher education projects to improve school mathematics (Stanic and Kilpatrick 1992, p. 413):

At first, the focus of curriculum development was the "college capable" student—the student who would be likely to enter college and who might be persuaded to pursue a scientific career if the school mathematics curriculum were more stimulating, intelligible, and mathematically elegant. During the early 1960s, several projects that had started by revising the secondary school curriculum began to tackle the elementary curriculum as well. Shortly thereafter, the federal government launched its War on Poverty, and the so-called disadvantaged student became a new focus of curriculum development work.... The elementary school proved a much tougher arena for the reformers, and ideas that appeared to work well with enthusiastic teachers and eager students in the high schools near universities often floundered when they were exported to less advantaged schools. Critics ... began to find a more receptive audience for their complaints that the new math was too abstract, impractical, and confusing. With the public, as well as educators themselves, increasingly convinced that the new math had failed, the wave of reform receded, and "Back to Basics" became the hallmark of textbooks and instructional programs.

In response to the back-to-basics movement of the 1970s, the National Council of Supervisors of Mathematics (NCSM 1978) issued a position paper on basic mathematical skills, and the NCTM (1980) followed with its *Agenda for Action*. These documents, rejecting any return to basics defined only as paper-and-pencil computation, called for a more balanced approach to school mathematics that would include problem solving, understanding, and applications to realistic situations.

The documents also set the stage for a more elaborate attempt by the NCTM to change curriculum, instruction, and assessment, the three key areas of school mathematics seen as needing reform. In 1989, the NCTM published its *Curriculum and Evaluation Standards for School Mathematics*. That was followed in 1991 by its *Professional Standards for Teaching Mathematics*. Its *Assessment Standards for School Mathematics* is being prepared for publication.

With these standards documents, the NCTM has entered "a new dimension in professional leadership" (Crosswhite 1990). Each document has been widely circulated in draft form and has received reviews and endorsements from a wide range of groups, including groups outside mathematics education. The final publications have been launched with considerable fanfare, accompanied by national press coverage and executive summaries for policymakers (for more on the process, see Crosswhite 1990; Massell 1993). The few criticisms of the documents have been muted (e.g., Effros 1990; Finn 1993; Palais 1989). Moreover, the NCTM has followed the documents with other activities and publications designed to help teachers understand and implement the

Standards. From its early days of being buffeted by other reformers and later trying to keep up with the pack, the NCTM has moved to the forefront of education reform.

The story of recent efforts to reform mathematics education, however, is neither as simple nor as straightforward as the foregoing sketch might suggest. It is not just a story of the NCTM's changed role. The efforts of the 1980s and 1990s are set within a context of other developments in education and in society. It would be premature to attempt a history of these recent reform efforts. Clearly, there are many exciting ideas around. What is not so clear is how school mathematics is changing as a consequence. The remainder of this essay is a personal view of recent efforts as refracted through the lenses of yesterday's lessons and today's contexts of reform.

LESSONS OF REFORM

Reform efforts in school mathematics have been marked by splits within and between the groups supporting or opposing the reform. These splits can be seen even in the rationale for establishing the NCTM. In the first issue of the NCTM's newly acquired *Mathematics Teacher*, the following "reasons for the National Council" were listed:

> First, it will at all times keep the interests of mathematics before the educational world. Instead of continual criticism at educational meetings, we intend to present constructive programs, by friends of mathematics. We prefer that curriculum studies and reforms and adjustments come from the teachers of mathematics rather than from the educational reformers.

> Second, it will furnish a medium through which teachers in one part of the country may know what is going on in every other part of the country. Significant reports and studies and experiments will be given wide circulation through the official journal. Otherwise they would be known only locally.

> Third, the Council through its journal will furnish a medium of expression for all of the teachers of the country. Thus, a feeling of solidarity will be aroused. All teachers of mathematics will know that they are members of one family, working together under a common leadership.

> Fourth, the Council will help the progressive teacher to be more progressive. It will also arouse the conservative teacher from his satisfaction and cause him to take a few steps ahead.

> Fifth, the splendid work of the National Committee on National Requirements will be conserved and extended as time and new conditions may show the need.

Sixth, the Council should receive the support of every teacher of secondary mathematics because it intends to maintain a journal whose editors and writers are real class room teachers. In fact, the teachers themselves are the owners of the journal, if they will but join the Council. (Austin 1921, p. 3)

The first and fourth reasons for establishing NCTM are particularly revealing. The organizers were, at one and the same time, leery of "educational reformers" and supportive of "progressive teachers." Given the diversity of suggested reforms inside and outside of mathematics education, these reasons, although not entirely clear, are not necessarily contradictory. It appears as though the NCTM defined progressives as supporters of mathematics and labeled their foes so-called educational reformers.

By the 1920s, after more than two decades of criticisms of the role of mathematics in the school curriculum, it was certainly not easy to define *friend* and *foe*, and it was even more difficult to label particular individuals (apart, perhaps, from unpopular critics such as David Snedden or Franklin Bobbitt) as one or the other. When the report of the National Committee on Mathematical Requirements (NCMR) appeared in 1923, it showed the effects on the committee's recommendations of the persistent attacks on mathematics that NCTM President Austin had noted in 1921. The NCMR, formed in 1916 and composed of some of the most vigorous defenders of mathematics, described its main goal as follows (p. 23):

> To the end that all pupils in the period of secondary education shall gain early a broad view of the whole field of elementary mathematics, and in particular, in order to insure contact with this important element in secondary education on the part of the very large number of pupils who, for one reason or another, drop out of school by the end of the ninth year, the National Committee recommends emphatically that the course of study in mathematics during the seventh, eighth, and ninth years contain the fundamental notions of arithmetic, of algebra, of intuitive geometry, of numerical trigonometry, and at least an introduction to demonstrative geometry, and that this body of material be required of all secondary pupils.

The proposed curriculum was indeed rigorous, but the concession to the critics was, in the context of the era, clear. The NCMR recognized that the battleground was shifting to the developing junior high school, and it was there that the committee members defended their stand for a reasonably rigorous mathematics program to be required of all students. Those students remaining in school after junior high were dealt with in the committee's discussion of "Elective Courses in Mathematics for Secondary Schools."

Advocates for less required mathematics even came from the community of teachers being asked to join the NCTM. Data from a survey of 416 secondary school teachers (forty-eight of whom taught mathematics) led Counts (1926) to conclude: "With the exception of the teachers of mathematics, practically all the teachers in these high schools feel that their own subjects should be more largely patronized than at present" (p. 156). Of the forty-eight mathematics teachers, seventeen thought that more pupils should take mathematics, thirteen thought that the same number of pupils should take mathematics, and eighteen thought that fewer pupils should take mathematics. The evidence, although certainly not overwhelming, suggests that not all the teachers called to join the NCTM wholeheartedly endorsed the cause. The other writings of the NCMR members reveal that their willingness to support less required mathematics (or to require mathematics only in the junior high school) was a grudging concession to more than twenty years of criticism. The teachers in the survey, on the other hand, had little reason to be anything other than completely honest about their beliefs.

The early twentieth century was, therefore, characterized by a complex school reform milieu. A variety of educators (including the forebears of present-day mathematics educators), sociologists, and others outside education were attempting to respond, in general, to changes in society and, in particular, to actual changes in the quantity and perceived changes in the quality of the school population. Ideas with very different origins and intentions began to come together under the general banner of reform. One of those ideas, unified mathematics, is a notable example of how bringing together diverse audiences through a common reform rhetoric can have problematic consequences.

Calls for a unified secondary mathematics curriculum had begun as early as 1890. Almost immediately, the proposals began to take different forms as they encountered opposition both inside and outside the mathematics education community (Sigurdson 1962, pp. 529–40). "The unified mathematics movement was not itself unified; significant differences of opinion existed among those people who were calling for mathematics to be organized into correlated, fused, unified, or parallel courses" (Stanic and Kilpatrick 1992, p. 410). Eventually, the unification movement began to focus on the junior high school years, where the standard curriculum was not so well entrenched. It also began to lose ground under the growing criticism of mathematics itself as a school subject. The student population of high schools, in particular, was expanding and changing in response to urbanization, industrialization, and immigration. "A rigorous unified mathematics curriculum for all students became, by the 1930s, the general mathematics taken by those students deemed incapable of doing higher level mathematics" (Stanic and Kilpatrick, p. 414).

At least one reason for that change was the approach taken by William David Reeve, a strong advocate of unified mathematics who was disappointed with its limited implementation in classrooms. While critics such as H. C.

Morrison, David Snedden, and Abraham Flexner were arguing that too much mathematics was required of students given its role in most people's daily lives, Reeve was trying to link his calls for unified mathematics to these broader criticisms.

Where other mathematics educators saw the work of reformers outside mathematics education as "severe criticisms against present practices in mathematics" (Minnick 1916, p. 81), Reeve saw an opportunity. Citing a critique by Morrison, he observed:

> No one who reads [it] will feel that he is attempting to discredit the place of mathematics as part of the child's educational equipment. On the other hand, his attitude is friendly and constructive—this, it seems, is one good reason why all mathematics teachers should read what he has to say. (Reeve 1916, p. 204)

Responding to Flexner's belief that "the course in mathematics should include 'nothing for which an affirmative case can not now be made out,'" Reeve replied, "This is exactly what teachers of mathematics who are really interested in the future are trying to do" (p. 207). Reeve recognized that he was linking a reform movement with its own history in mathematics education with the reform rhetoric of the "outsiders":

> Let it be remembered that these suggestions . . . are not given merely in the hope that something may be done to satisfy the cry for reform in the teaching of mathematics (which is in many respects legitimate) but to try to remedy some of the weaknesses of the system of which we were aware long ago, and also to increase the mathematical power of the pupils who go out from our classes so that their knowledge can function most efficiently. (pp. 211–212)

Reeve continued his attempts to use the banner of general school reform to advocate unified mathematics with a keen awareness of his audiences. For example, in 1920 he addressed members of the NCTM on the topic of a "general" mathematics program by saying, "I am not interested in a destructive type of criticism of past methods" (Reeve 1922, p. 381). Speaking to principals and superintendents of the Minnesota Educational Association that same year, he gave essentially the same address, but with a different spin. He began with a greater willingness to recognize past criticisms:

> For a long time, mathematicians and others have occasionally denounced our methods of teaching mathematics and usually from their own point of view.... A careful study of the recent tirades against mathematics reveals criticisms really aimed at faulty organization and improper methods of teaching the subject matter, rather than a desire to discredit the subject itself. (Reeve 1920, p. 258)

He went on to associate himself with the reformers' view that the study of mathematics cannot be justified on the basis of tradition alone but must be shown to have an appropriate role in students' lives.

Clearly, Reeve was adjusting his message. On the one hand, the strategy is not unreasonable. Anyone who wants a message to be accepted must consider the audience. On the other hand, given Reeve's (1920) desire for a program that would lead all students into "much higher and more powerful mathematics without any ultimate loss" (p. 260), his intention does not seem to have meshed well with the reform rhetoric he was willing to use.

Others attempted to adapt their calls for unified mathematics in a similar fashion (see, e.g., Breslich 1916, 1920), but Reeve's role was especially significant given his position as editor of the *Mathematics Teacher* during the 1930s. His editorials announcing the crisis in school mathematics and defending the role of mathematics in the curriculum must have been written with some regret that he had been too quick to accept the language and rhetoric of the critics.

Reeve was far from the last mathematics educator to pick up the banner of school reform as a means of legitimizing proposed changes in school mathematics. The principal recommendation in the *Agenda for Action* (NCTM 1980) was that "problem solving be the focus of school mathematics" (p. 1). By characterizing problem solving as a basic skill, the authors of the *Agenda* appropriated, reinterpreted, and hoped to turn to their own advantage the language of the back-to-basics reformers. Essentially, the argument was that the *Agenda* authors, too, supported "going back to the basics," as long as everything the NCTM wanted in a mathematics curriculum was defined as a basic skill.

The strategy, however, had its drawbacks. Placing the problem-solving recommendation within the context of the back-to-basics rhetoric both distorted the nature of problem solving as an approach to teaching mathematics and implicitly supported those with a more limited view of school mathematics. In any school subject, but perhaps especially in mathematics, *how* skills are taught is as critical as *which* ones are taught. Students need to see skills not as ends in themselves but as means to other ends. As John Dewey (1910) said, there is "danger in those studies where the main emphasis is upon the acquisition of skill.... Practical skill, modes of effective technique, can be intelligently, non-mechanically *used* only when intelligence has played a part in their *acquisition*" (pp. 51–2). Although the authors of *An Agenda for Action* would be likely to agree with Dewey, their call for problem solving to be the focus of school mathematics was undermined by their willingness to take what might be seen as a general approach to teaching mathematics and reduce it to one of many skills to be learned. There is a price to be paid when slogans are stretched to cover multiple agendas.

A NATION AT RISK

The publication *A Nation at Risk* (National Commission on Excellence in Education 1983), which is usually seen as launching today's reform movement in education, actually appeared after a number of other activities associated with reform inside and outside mathematics education were under way. Inside mathematics education, efforts were being made to update the curriculum, incorporating more applications and giving technology a greater role in mathematics instruction (see, e.g., College Board 1983; Conference Board of the Mathematical Sciences 1982; NCSM 1978; NCTM 1980; National Science Board Commission on Precollege Education in Mathematics, Science, and Technology 1982; and National Science Foundation and U.S. Department of Education 1980). Meanwhile, a larger movement to reform education nationwide had begun as well (Finn and Rebarber 1992, pp. 175–6),

> perhaps with the 1975 revelation by the College Board that average Scholastic Aptitude Test scores had been declining for a decade, with media attention to "illiterate" high school graduates and with the "back to basics" movement that began in the late 1970s.

The test-score decline issue grabbed the public's attention and reinforced the use of test scores as measures of the quality of schools and schooling. The back-to-basics movement, which for many parents and other citizens has never abated, was accompanied by calls that schools (and teachers) be held accountable for the product they were supposed to be turning out—educated students.

The *Nation at Risk* document, however, was to set many of the terms of the discourse, including the economic competitiveness arguments for reform and the justification of reform in the low performance of students on international assessments. The tone of *A Nation at Risk* (p. 5) was strident and certain:

> Our Nation is at risk. Our once unchallenged preeminence in commerce, industry, science, and technological innovation is being overtaken by competitors throughout the world.... The educational foundations of our society are presently being eroded by a rising tide of mediocrity that threatens our very future as a Nation and a people.

Today, the report's bleak view of the United States as a potentially second-rate world power has given way to more optimistic appraisals. Germany, South Korea, and Japan—countries cited in the report as threats to U.S. prowess—have been mired in their own economic problems. The Soviet Union, along with the military threat it posed, has disappeared. Though America's "position in the world" is not without its problems, the country's position is secure in ways not anticipated by the report.

Despite these changes, it is much too soon for any effects of the current wave of education reform to have been felt in the economy. Any improvements we see cannot be attributed to changes in education. *A Nation at Risk* assumed that the educational and the economic systems in a country are so linked that improvement in the former inexorably improves the latter. That assumption has always been questionable. Nonetheless, it was embraced at the time by reformers in various camps, who recognized that more subtle arguments for improving education might not capture public attention. Today's reform rhetoric is still set within a discourse of failure in which American schools are portrayed as having failed in their mission, American teachers as having failed to teach, and American schoolchildren as having failed to learn. The perception that U.S. students are failing to keep ahead in the international race has turned out to be a powerful metaphor for justifying educational reform.

Setting the argument for reform in terms of test scores, however, simply fueled efforts to seek accountability in schools by testing students. State governments, having increased their expenditures on education, were expanding their testing programs as a means of controlling what the schools were doing with those expenditures. With many more testing programs in place, the stage was set for policymakers arguing for reform to present charts comparing average scores of countries, states, and districts, as well as graphs showing fluctuations in scores. The charts and graphs could be used not only to show that the system was failing but to detect any indications of improvement.

The focus on accountability was complicated by various movements within education to restructure schools and to give more autonomy to teachers (Carnegie Task Force on Teaching as a Profession 1986; Darling-Hammond 1990; Finn and Rebarber 1992). Teachers were then caught between those reformers calling for more accountability (e.g., higher scores by students on mandated tests) and those reformers calling for more autonomy (e.g., greater reliance on teacher judgments of student progress). Because many testing programs work at odds with other reform ideas, assessment reform has become a critical battleground.

Within mathematics education, reform was helped along by societal concerns about inequitable access to professions involving mathematics, science, and technology. The theme of "access to knowledge" (Goodlad and Keating 1990) was sounded, and questions were raised anew about the effects of practices such as tracking and ability grouping in mathematics. The content of the mathematics curriculum itself came into question as researchers demonstrated the disparity between school mathematics and the mathematics used in society (e.g., Rogoff and Lave 1984). Some reformers of mathematics education proposed that the subject could be made more relevant to groups traditionally underrepresented in mathematics, and thereby presumably more comprehensible to all, if it were culture-inclusive (Wilson and Mosquera Padron 1994). Other reformers worked on bringing more technology into the mathematics curriculum and into instruction. All the reformers apparently saw a need to turn away from the abstract mathematics that was prominent during the new math era toward a more applied mathematics that emphasized data analysis and mathematical modeling. Finding a balance between abstraction and application, however, is a dilemma the mathematics education community must still resolve.

RAISING STANDARDS

Undoubtedly the most adroit move—one that was perhaps also the most problematic—by those who sought reform in school mathematics a decade or so ago was the appropriation of the term *standards* to describe the proposed changes. The term is irresistible to politicians and the public, who see it as restoring lost qualities to schooling, raising expectations as to what students will learn.

A key recommendation in *A Nation at Risk* referred to "Standards and Expectations":

> **We recommend** that schools, colleges, and universities adopt more rigorous and measurable standards, and higher expectations, for academic performance and student conduct, and that 4-year colleges and universities raise their requirements for admission. This will help students do their best educationally with challenging materials in an environment that supports learning and authentic accomplishment. (p. 27)

Little in the subsequent prose about implementing the recommendation referred to the curriculum or to the process of instruction. The emphasis, instead, was on changes in such indicators as grading practices, admissions requirements, and standardized test scores to show that higher standards and expectations had been met. Textbooks and instructional materials were to be upgraded and updated, but that hardly constitutes reform.

The reformers of school mathematics no doubt wanted to raise expectations, but more than that they wanted to change the content of school mathematics, how it would be taught, and how it would be assessed. The NCTM *Curriculum and Evaluation Standards* (1989) proposed three reasons for groups to adopt a formal set of standards: "(1) to ensure quality, (2) to indicate goals, and (3) to promote change" (p. 2). By using the language of standards, the NCTM could lay out its goals and its hopes for change in a form that would speak to the profession about a vision for school mathematics and to the politicians and public about improved learning. Even a cursory reading of the reaction to the NCTM Standards by educators and the public, however, suggests that improved performance, shown primarily by higher test scores, is what a standards document is expected to address.

For those inside the reform community—NCTM members and certain other professionals—the Standards are a vision of what could be, a flag to rally around, but clearly not a road map with detailed indicators of progress. For the more diverse community outside comprising the great mass of people teaching, creating, or using mathematics, as well as policymakers and the public at large, the Standards by their very name must provide benchmarks of quality and means of determining improvement. Once again, this time by appropriating the rhetoric of standards, the leaders of reform in mathematics education have found themselves caught in a web of language.

THE RHETORIC OF REFORM

Our contention is that *reform* is too strong a word to characterize developments in mathematics education at any point over the past century. Professional and public discussion of issues in mathematics education ebbs and flows. School mathematics continually changes, but it has yet to achieve a form substantially different from that being established in the closing years of the last century. Change should not be confused with reform. Only two historical moments—the decades at the turn of the century, when a more unified, applied curriculum was proposed, and the 1950s and 1960s, when the curriculum was to be modernized—come close to qualifying as times of reform. Yet both yielded only surface changes in school mathematics, having effects quite different from the reformers' intentions (Stanic and Kilpatrick 1992).

What has happened during the past century is that the rhetoric of reform, always inflated to grab attention, has been simplified, made even more dramatic, and brought onto the national stage. Meanwhile, the views and actions of teachers and students of mathematics have remained richly diverse. Reform movements have a kind of enforced unity at the top, but disunity always breaks out below. The accompanying rhetoric tends to paper over that disunity.

In 1920, when the NCTM was founded, the subject of mathematics was endangered in the curriculum. Three-quarters of a century later, the public school is endangered in the sense of being rejected by the public in favor of various alternatives. Also endangered, sometimes literally, are the people within public schools. The reform movement in mathematics education underestimates this wider social context while taking advantage of the public's and the policymakers' desires for reform. The rhetoric plays on anxieties about changes in the workforce but ignores students' declining interest in education as a form of advancement and their legitimate concerns about limited employment opportunities. It also assumes a role for teachers that in many cases conflicts with their daily experience. In short, the wider context is recognized superficially as a source of justification for reform but is neglected, often ignored, for its impact on the implementation of reform ideas.

Today's reformers of mathematics education may have learned the lessons to be drawn from previous reform efforts. They almost certainly recognize that true reform has not occurred up to now, and they know that it cannot be brought about simply by preaching—either to the already converted or to those unfamiliar with the substance of the reform. One can argue that today's reformers are better at doing what our forebears consistently attempted, namely, to build a community of vision. True reform, unfortunately, may require doing not something better but something different. Given our own limited perception of a different path to take, we can only applaud the persistence of those pursuing a vision of mathematics education that, although popular, is neither new nor easily implemented.

· · · · · · · · · · · · · · ·

NOTE

Our thanks to Douglas McLeod for providing some of the references.

· · · · · · · · · · · · · · ·

BIBLIOGRAPHY

Austin, C. M. "The National Council of Teachers of Mathematics." *Mathematics Teacher* 14 (1921): 1–4.

_____. "Historical Account of the Origin and Growth of the National Council of Teachers of Mathematics." *Mathematics Teacher* 21 (1928): 204–13.

Breslich, E. R. "Correlation of Mathematical Subjects." *School Science and Mathematics* 20 (1920): 125–34.

_____. "Forward Movements in Secondary Mathematics." *School Review* 24 (1916): 283–97.

Carnegie Task Force on Teaching as a Profession. *A Nation Prepared: Teachers for the 21st Century.* New York: Carnegie Forum on Education and the Economy, 1986.

College Board. *Academic Preparation for College: What Students Need to Know and Be Able to Do.* New York: College Board, 1983.

Conference Board of the Mathematical Sciences. *The Mathematical Sciences Curriculum K–12: What Is Still Fundamental and What Is Not.* Washington, D.C.: The Board, 1982.

Counts, G. S. "Current Practices in Curriculum-Making in Public High School." In *Curriculum-Making: Past and Present,* Twenty-sixth Yearbook of the National

Society for the Study of Education, edited by G. M. Whipple, Pt. 1, pp. 135–62. Bloomington, Ill.: Public School Publishing, 1926.

Crosswhite, F. Joe. "National Standards: A New Dimension in Professional Leadership." *School Science and Mathematics* 90 (1990): 454–66.

Darling-Hammond, Linda. "Achieving Our Goals: Superficial or Structural Reforms?" *Phi Delta Kappan* 72 (1990): 286–95.

Dewey, John. *How We Think.* Boston: Heath, 1910.

Effros, E. G. "Some Thoughts on `Everybody Counts.'" *Notices of the American Mathematical Society* 37 (1990): 555–61.

Finn, C. E., Jr. "What If Those Math Standards Are Wrong?" *Education Week,* 20 January 1993, pp. 36, 26.

Finn, C. E., Jr., and Theodor Rebarber. "The Changing Politics of Education Reform." In *Education Reform in the '90s,* edited by C. E. Finn, Jr., and Theodor Rebarber, pp. 175–93. New York: Macmillan, 1992.

Goodlad, John I., and Pamela Keating, eds. *Access to Knowledge: An Agenda for Our Nation's Schools.* New York: College Board, 1990.

Kilpatrick, Jeremy. "A History of Research in Mathematics Education." In *Handbook of Research on Mathematics Teaching and Learning,* edited by Douglas A. Grouws, pp. 3–38. New York: Macmillan, 1992.

Massell, Diane. "Setting Standards in Mathematics and Social Studies." *Education and Urban Society* 26 (1994): 118–40.

Minnick, J. H. "Our Critics and Their Viewpoints." *Mathematics Teacher* 9 (1916): 80–4.

National Commission on Excellence in Education. *A Nation at Risk: The Imperative for Educational Reform.* Washington, D.C.: U.S. Government Printing Office, 1983.

National Committee on Mathematical Requirements. *The Reorganization of Mathematics in Secondary Education.* Washington, D.C.: Mathematical Association of America, 1923.

National Council of Supervisors of Mathematics. "Position Statements on Basic Skills." *Mathematics Teacher* 71 (1978): 147–52.

National Council of Teachers of Mathematics. *An Agenda for Action: Recommendations for School Mathematics of the 1980s.* Reston, Va.: The Council, 1980.

_____. *Assessment Standards for School Mathematics.* Reston, Va.: The Council, forthcoming.

_____. *Curriculum and Evaluation Standards for School Mathematics.* Reston, Va.: The Council, 1989.

_____. *Professional Standards for Teaching Mathematics.* Reston, Va.: The Council, 1991.

National Science Board Commission on Precollege Education in Mathematics, Science, and Technology. *Today's Problems, Tomorrow's Crises.* Washington, D.C.: National Science Foundation, 1982.

National Science Foundation and U.S. Department of Education. *Science and Engineering Education for the 1980's and Beyond.* Washington, D.C.: National Science Foundation, 1980.

Osborne, Alan R., and F. Joe Crosswhite. "Forces and Issues Related to Curriculum and Instruction, 7–12." In *A History of Mathematics Education in the United States and Canada*, Thirty-second Yearbook of the National Council of Teachers of Mathematics, edited by Phillip S. Jones, pp. 155–297. Washington, D.C.: The Council, 1970.

Palais, E. G. "A Differing View on Mathematics Education Reform." *Notices of the American Mathematical Society* 36 (1989): 1889–991.

Reeve, W. D. "The Case for General Mathematics." *Mathematics Teacher* 15 (1922): 381–91.

_____. "General Mathematics for the High School: Its Purpose and Content." *Educational Administration and Supervision* 6 (1920): 258–73.

_____. "Unification of Mathematics in the High School." *School and Society* 4 (1916): 203–12.

Rogoff, Barbara, and Jean Lave, eds. *Everyday Cognition: Its Development in Social Context.* Cambridge: Harvard University Press, 1984.

Sigurdson, S. E. "The Development of the Idea of Unified Mathematics in the Secondary School Curriculum: 1890–1930." Doctoral dissertation, University of Wisconsin at Madison, 1962. *Dissertation Abstracts International* 23 (1962): 1997.

Smith, David Eugene. "Among My Autographs, 26: Burckhardt on Modern Teaching." *American Mathematical Monthly* 29 (1922): 297–9.

Stanic, George M. A. "The Growing Crisis in Mathematics Education in the Early Twentieth Century." *Journal for Research in Mathematics Education* 17 (1986): 190–205.

Stanic, George M. A., and Jeremy Kilpatrick. "Mathematics Curriculum Reform in the United States: A Historical Perspective." *International Journal of Educational Research* 5 (1992): 407–17.

Wilson, P. S., and J. C. Mosquera Padron. "Moving toward Culture-Inclusive Mathematics Education." In *Multicultural Education: Inclusion of All*, edited by M. M. Atwater, Kelly Radzik-Marsh, and Marilyn Strutchens, pp. 39–63. Athens, Ga.: University of Georgia, Department of Science Education, 1994.

STUDENTS . . .

As we approach the next century, mathematics teachers are confronted with the challenge of preparing all students with the mathematical-reasoning, communication, and critical-thinking skills that will enable them to become responsible citizens. However, a proliferation of research documents underachievement in mathematics by certain cultural, racial, gender, and socioeconomic groups. In this section, mathematics educators, administrators, and researchers present provocative ideas, new strategies, and policy changes that support the belief that every student must have an equal opportunity to develop problem-solving skills and high-level mathematical reasoning. The papers included in this section argue a convincing case for why instruction in mathematics, at all levels, must be organized in a way that accommodates the differences in children's learning styles, dispositions, and experiential backgrounds. An abundance of teaching ideas that can be used to develop students' reasoning and communication skills, using novel problems and applications based on real-life situations, are presented to demonstrate that all students, when provided with optimal teaching and learning conditions, are able to achieve high standards.

Several authors call for the reshaping of mathematics education so that all students are helped to achieve their full potential. This section also presents a detailed exposition of some of the possible causes of differences in achievement in mathematics, including biases in test construction and the lack of commitment to gender equity.

—Lucille Croom

How Primary Students Think and Learn

Patricia F. Campbell & Martin L. Johnson

THROUGHOUT the history of education, mathematical literacy has always been viewed as essential as a foundation for democracy and as a profound and powerful part of human culture. The centrality of mathematics to human culture is reflected in the fact that mathematics is one of the few subjects studied by children throughout all of their years of elementary schooling and for many years beyond. Although many societal institutions share the responsibility to help each child develop mathematical knowledge, the major expectation for such development is placed on the schools.

SCHOOLS, CHILDREN, AND MATHEMATICAL LEARNING

The elementary school is the setting where young children confront most of their formal mathematical knowledge. As such, decisions about the mathematics curriculum and how the school will be organized for mathematics instruction have a critical impact on the mathematical knowledge base children will develop.

Schools are complex entities and contain characteristics of both social systems and cultural systems. As a social system, a school is an institution involving people having different and interrelated roles, statuses, and

Patricia Campbell is an associate professor in the Center for Mathematics Education at the University of Maryland at College Park. In addition to teaching graduate and undergraduate courses, she is conducting research on the teaching and learning of mathematics in predominantly minority, urban elementary schools.

Martin Johnson is a professor of mathematics education and chair of the Department of Curriculum and Instruction at the University of Maryland at College Park. He has taught courses at the undergraduate and graduate levels, conducted research, and written extensively on the mathematics learning of minority children.

responsibilities. People within this system function through patterns of actions and interactions. Schools contain classrooms with teachers and students where action and interactions are played out. In mathematics classrooms, the traditional teaching approach places the teacher as the source of mathematics knowledge. The teacher in this classroom usually asks the questions, presents the problems, shows how to solve a problem, and then asks students to use this model or demonstration to do other problems.

Most children in these classrooms become socialized into believing that the teacher knows how to solve every problem and hence the teacher has mathematical knowledge. Unfortunately, many of these children also infer that they cannot or need not solve a problem until a teacher or some other adult tells them how to solve it. Further, students come to believe that only persons who have mathematics knowledge and who can solve problems have mathematical power. Such an approach has permeated mathematics teaching for the major part of the twentieth century.

Schools may also be viewed as cultural systems, with sets of values, norms, ethoses, and shared meanings. Administrators and teachers make decisions that reflect their views on each of these dimensions as they decide on (a) the mathematics content of the school's curriculum, (b) the organization of the mathematics curriculum, (c) the instructional methodologies chosen, and (d) the opportunities provided for all students to learn important and useful mathematics.

Throughout history, schools have always attempted to ensure that students develop the mathematical knowledge needed to function in the workplace at that time. This has placed an enormous responsibility on schools to be current, yet futuristic, in their curriculum and methods. Determining what the mathematics curriculum should be and how instruction should be carried out have proved to be difficult decisions and have been the subject of much discussion over the decades. Every decade has recorded its successes and failures with the overall view that, in general, schools have failed at preparing students at large in mathematics.

Constraints on Student Success

Differential performance among students can often be associated with the constraints placed on them from many sources. These sources include the organizational structures of the school, the curriculum decisions made by administrators and teachers, and the basic beliefs of teachers and administrators regarding who should learn what mathematics. Clearly children come to school with different background experiences and with different dispositions to learn. Schools for primary children should allow for these differences and develop programs that allow students to enter into exciting mathematics from whatever background they come.

This requires a new look at the traditional, already "set in place" curriculum. Curricula must allow for multiple entry points so that more children can have the opportunity to build mathematical competence from whatever base they possess. Most current mathematics programs still present a list of prerequisite skills, and when strictly interpreted by a teacher, the result is often one group of students who continue to move forward and others who cycle into a remedial learning track. Organizational structures within the school, coupled with teacher beliefs about who can learn mathematics, have resulted in a very different curriculum for different groups of students (Oakes 1985; McKnight et al. 1987).

Some reports indicate that decisions about the type of mathematics curriculum, in particular, skills based versus problem-solving based, are often a function of the racial composition and economic level of the student body and the community from which the students come (Jackson 1990). Schools must be reorganized, curriculum must be rethought, and *teachers must become more concerned about insuring that all students achieve in mathematics at a very high level.*

Central to accomplishing the goal above is a new look at how children actually "construct" mathematical meanings (cognitive considerations) and the importance of social environments that support and encourage meaningful mathematics learning (social considerations). The following sections of this paper will discuss the types of interactions one would see in classrooms guided by a constructivist philosophy of teaching and learning.

Examples of children's thinking and of classroom events have been gathered as part of the National Science Foundation–funded Project IMPACT (Increasing the Mathematical Power of All Children and Teachers). This project is an effort to design, implement, and evaluate a model for elementary mathematics instruction that is compatible with the NCTM *Curriculum and Evaluation Standards for School Mathematics* (National Council of Teachers of Mathematics 1989) in predominantly minority, urban schools (Campbell and Rowan, forthcoming).

COGNITIVE CONSIDERATIONS

Construction of Mathematical Understanding

"One of the most widely accepted ideas within the mathematics education community is the idea that students should understand mathematics" (Hiebert and Carpenter 1992, p. 65). While the intent of mathematics instruction in the past may have been to stimulate recall of mathematical skills and rules, the goal today has advanced to mathematical understanding.

In particular, today a perspective that is becoming more prevalent is the presumption that children must "construct their own mathematical understanding" (National Research Council 1989, p. 58). The constructivist framework has been cited by researchers, curriculum developers, and administrators, so it may seem as if everyone is claiming a common perspective. Yet many of these proponents do not clarify exactly what the constructivist perspective means to them. Further, the implications of constructivism for application in the classroom generally lead advocates to adopt slightly differing interpretations. Consider, then, just what does it mean for young children to "construct understanding"?

Construct relationships

To know mathematics is to construct relationships that give order to, and define patterns across, experience. These relationships may be based on the similarities and differences that the child recognizes. For example, when a young child is learning about numbers, many teachers will have the child build a set ("Give me five chips, please") or count a set ("How many chips are there?"). But to truly develop a sense of what a number means, the child needs to experience the numbers under study in differing settings with different physical representations, always searching for commonality, for an ordered pattern. The teacher may tell story problems where it is necessary to count and make sets or to reflect on the relationships between numbers in order to answer the questions that occur in the problems. For example, consider the following exchange in a second-grade classroom. The children are offering their approaches for solving the following problem about a rabbit that was visiting the classroom: *Sniffles had 52 baby rabbits. 46 of the baby rabbits were girls. How many were boys?*

> *Shanita:* You take away.
>
> *Teacher:* Why do you say it's a "take away" problem?
>
> *Shanita:* Because you take away how much the girls are and you see how much the boys are.
>
> *Teacher:* Okay, so she would take away the number of girls and then she says that the left over would be the number of boys.... Does anyone else have another way to solve the problem? ... Miko?
>
> *Miko:* I don't think it's take away.
>
> *Teacher:* Okay, how would you solve the problem?

Miko: If I had 52 baby rabbits and 46 of the rabbits were girls, then.... How many boys there are.... So I would say add.

Teacher: How would you add it?

Miko: I would think of 46 girl rabbits and then go up, think of how many rabbits are boys. So that's like addition.

Teacher: So you would be counting up to the number? (Miko nods her head.) What number would you be counting up to?

Miko: 52.

Teacher: So you would start from this number (points to 46 in the problem) and then count up to 52? And then what would you do?

Miko: And then, then I'd see, I'd see that it's addition and I knew that the boys.... There were 6 boys.

Teacher: So you did this in your head.... Monique?

Monique: Well ... it's not a plus or a minus.

Teacher: It's not a plus or a minus? ... What kind of a problem is it then?

Monique: It's just asking another question.... You're not adding. You're not taking away anything.

Teacher: So, so how would you solve this problem then?

Monique: Just take the number 46 ... and then you'll count, then I would count to get some more. How many more to get 52. And then I would, I'll take away the 46. And then I'll see how many were left. So that's a take away problem.... And then you could add on 46 to 52.

Teacher: So you said it could kind of be both ways?

Monique: (nods her head)

Teacher: You're saying that it could be plus because we could count from 46 up to 52. Or we could have 52 all together and take away the 46. To you, that's what the problem is?

Monique: (nods her head)

Over time, number, addition, and subtraction will gradually be recognized and constructed by the child. In time, the child comes to understand that number and numerical relationships exist distinct from their representation, that numerical relationships occur in many settings in everyday events within the child's environment, that physical representations of quantities as well as relationships between quantities can be investigated by counting, that symbols can be written for numbers and relationships, and that numbers can be used to compose and decompose other numbers.

To illustrate better the interaction of these emerging understandings, consider a second-grade boy's solution to the following problem: *Anna has 46 stickers. Her sticker book holds 75 stickers. How many more stickers does she need to fill her sticker book?* Beneath the problem, the handwritten statement $46 + \underline{\quad} = 75$ has been written. The child has been told that someone else had started to work this problem and wrote down what he was going to work on, but then that person did not finish it. The child has been told that he can use that person's writing if he wants to, but that he does not have to use it. The child silently rereads the problem, thinks for a number of seconds, and then says, "I think my answer is 29, but I'm going to add 29 plus 46 ... to get my answer." The child then adds $46 + 29$ vertically using the traditional algorithm to obtain the solution 75. He then writes 29 on the blank line in the handwritten number sentence.

> *Teacher:* Now you came up with the answer 29, and this was your checking, right? (pointing to the traditional addition problem)
>
> *Tyrone:* Uh huh. (nods his head)
>
> *Teacher:* How did you get the answer 29 in the first place?
>
> *Tyrone:* Well ... 6 is more than 5, right?
>
> *Teacher:* Yeah. (stated in a puzzled fashion)
>
> *Tyrone:* Well, you're going to need a 9 to get a 5. And 6 plus 9 is 15? ... Right, so that's how I got the 5.... I don't know how I really did it. See.... No ... I know now.... 20, 20 plus 40, 'cause this stands for a ten (points to the 4 in $46 + \underline{29} = 75$). 20 plus 40 is 66, right? Then, then you, then I added a 9, plus the 20 is 29.
>
> *Teacher:* Ohhh.
>
> *Tyrone:* Because ... see, 6 is more than 5. If this was a 5 (points to the 6 in $46 + \underline{29} = 75$), then I would have picked 30 because ... 10 plus 5 is 15. 9 plus ... 9 plus 6 is 15. I need to get the 5 right there (points to the 5 in $46 + \underline{29} = 75$). Then I got the whole answer.

Another aspect of constructing relationships is to perceive a situation as a component of, or as being included in, a "big idea." Rather than viewing mathematics as a colossal collection of disconnected terms, rules, and procedures, mathematics can be viewed as a smaller set of interrelated big ideas. To understand mathematics then means to construct meaning for these big ideas. For example, traditionally addition and subtraction have been taught to young children in isolation, perhaps at first with sums limited to 10 or less and then to sums to 18 or 20.

The rationale for this was simple. There were too many pieces of information to recall if all of the individual addition and subtraction facts summing to 18 or 20 were to be learned simultaneously. Now, however, researchers have learned that children may come to construct the meaning of addition and subtraction through meaningful word problems (Carpenter and Moser 1984). The word problems provide a setting for the developing concepts of addition and subtraction and may foster strategies for these operations whether they are presented in computation or story-problem formats.

Rather than deferring word problems until after a child has conquered computation, research suggests that children may develop addition and subtraction strategies by solving word problems. How can children solve a word problem if they have never learned their basic facts? Children can model the quantities or abstract the numerical relationships in the problem to reach a solution. The following solutions were offered by two first-grade children at the end of the academic year. They were both solving the problem: *Lateesha has 15 baseball cards. David has 9 baseball cards. How many more baseball cards does Lateesha have than David?*

Marcus: Lateesha has how many, 15?

Teacher: Fifteen. I'll tell you the whole story again, okay? Lateesha has 15 baseball cards. David has 9 baseball cards. How many more baseball cards does Lateesha have than David?

Marcus: Six.

Teacher: How did you do that?

Marcus: You have to, you have to start from 9 and count on to, count on to 15 and keep track of what, of what you counted.

Teacher: Lateesha has 15 baseball cards. David has 9 baseball cards. How many more baseball cards does Lateesha have than David?

Justine: Six.

Teacher: Now how did you do that?

Justine: Well ... I thought, um, 15 from 6. Ten plus 5 is 15, and 9 is one less and.... You, you should put one more for the, for the 10.

Eventually children may abstract addition and subtraction events as a part-whole relationship (Riley, Greeno, and Heller 1983). Once a child has constructed the part-total or part-whole relationship, the child can think of number, addition, and subtraction as differing conditions that compose one big idea. A number is a total, but it can be thought of as being made up of other numbers, its parts. In addition, two parts are known, and they yield a total. In subtraction, the total and one part of that quantity are known; the intent is to determine the other part that composes the total. Indeed, "probably the major conceptual achievement of the early school years is the interpretation of numbers in terms of part and whole relationships" (Resnick 1983, p. 114).

The power of the part-whole abstraction is evident when children explain their solution strategies. In the following excerpt, a first-grade boy has been given the written stimulus 68 – 50 = ___. In his class, he has begun to investigate the meaning of two-digit numbers, taking physical quantities up to 100, separating the objects into sets of ten and leftover ones, and recounting the total. He has also solved addition and subtraction word problems and written number sentences up to a sum of 18. At the time he was given this problem, this child had never before seen two-digit numbers written within an addition or subtraction number sentence. There were Unifix cubes available for the child to use, but he did not do so.

> *Damian:* (Thinks for 22 seconds and then writes 18.)
>
> *Teacher:* Okay, Damian, tell me how you solved that problem.
>
> *Damian:* Fifty plus 50 ... I mean, 50 minus 50 is zero, right?
>
> *Teacher:* Uh huh.
>
> *Damian:* And then, this has another 10 to add to it (pointing to the 68). So, so there would be 10 left. But there's an 8, so you add the 10 plus 8 is 18. So there's 18 left.
>
> *Teacher:* Thank you, Damian.

In the next excerpt, a third-grade girl has been given the problem 74 – 26 to solve as she thinks aloud. She, too, abstracts a part-whole relationship as she finds parts that are multiples of 10 in numbers, separating the multiples or breaking them apart as needed.

> *Laura:* Okay, I take away the ... 20 from the 70 and I have 50.... No, from the 74 and I have 54. And then I break up the ... one of the tens and now I have 14 ones. And, I count out 6 of those and I have 8 ones left. And ... it's 48.

Active involvement

No one can give a child mathematical knowledge. It cannot be "transmitted" or "absorbed" or "received." Developing mathematical knowledge is not a matter of memorizing rules or practicing skills. Mathematical knowledge comes from an individual's effort to connect new experiences or conclusions to prior knowledge. These connections must be constructed by each child. It is not simply a matter of a child "paying attention to what is being said" by the teacher in school. Rather, it is a matter of the child actively attempting to construct and test the meaning of new experiences in light of prior understanding.

The most difficult aspect of the constructivist perspective to apply in classrooms is the implication of the statement "Each child must *construct* his or her mathematical understanding." For many teachers the issue is, "How will they [the children] know the mathematics if we don't tell them?"

Consider the experience of one first-grade teacher. She was willing to support active, hands-on experiences in her classroom using many manipulatives within child-centered activities. However, even after two years of participating in Project IMPACT, she admitted that she still had to monitor her impulse to intervene when children encountered complex mathematics problems. She likes to recount this classroom experience as evidence of what children will attempt, and sometimes attain, when they are truly interested in a problem.

> We were doing fractions and I was having the kids dividing different objects into parts with masking tape. The children got to choose the object they wanted to divide and then draw a number from a pile. This number told them how many parts they were to create. One group of three average-achieving girls chose to divide the "cubbies," an area of thirty-two separate boxes stacked sixteen on top of sixteen. The number they drew was six. [Cubbies are stacked wooden boxes affixed on one classroom wall about three feet off the ground. The boxes open toward the classroom. This is where the children keep their backpacks, papers, and lunchboxes.]

> One-sixth of thirty-two! Well, that was one of my most difficult problems of the day, certainly more difficult than any I'd envisioned when planning this activity. When I saw the task ahead of them I thought, "They'll never be able to do this." But they wanted to try. So I went over to them and started asking a series of questions, trying to sway them to do it my way. But they wouldn't hear of it, saying, "No, no, we've got our own ideas," so I went on to other children.

> From time to time I would look over toward the cubbies and notice that things looked confused. They had tape all over. When it came to their turn to talk about what they'd done, I thought, "I'll skip them, maybe they won't notice." But no, they had to explain. I thought, "Oh, these poor girls, they're not going to be able to explain this."

> Then one of the girls said, "One-sixth of thirty-two cubbies is five and one-third." All I could say was, "WHAT?" and she said it again, "One-sixth of thirty-two cubbies is five and one-third." And I said, "Can you explain that?"

> The girls explained how they took the cubbies and divided them into six parts and they had five cubbies in each part. They had two cubbies left over,

so they took those two cubbies and divided each one into thirds, so that would be six. Then they went back and counted and it was five and one-third.

Well, I was just shocked with what they were doing! If you wrote on a piece of paper one-sixth of thirty-two, it would seem an impossible task for first graders. But the cubbies were there, they had the tape, they could count, they could change it around. (Rowan and Bourne 1994, pp. 27–28)

Children can answer questions about mathematical experiences even if they have not been told the answers. The constructivist perspective posits that children learn as they attempt to solve meaningful problems, with manipulatives or other necessary materials available should the children decide to use them. The teacher's role is to ask appropriate questions to provoke the children to reflect on their actions and to validate their approaches and reasoning.

The questioning is critical. Without the questioning it may be possible for a child to complete an activity or solve a problem, manipulate the objects, obtain a "correct" solution, and never confront the mathematics that the teacher intended the activity or problem to foster. It is the reflection catalyzed by the questioning that may lead to the child's assimilation, accommodation, or integration of new understandings.

Learning in the Classroom

This approach is not discovery learning. It is not a matter of the teacher or the curriculum developer breaking a problem or task into substeps that are readily solved by the child and then presuming, because the child accomplished each of the obvious substeps, that the child has solved the problem. This approach is also not posing a problem, with necessary materials present, and then waiting for a miracle to happen.

When situated in the reality of the classroom, the constructivist perspective demands that teachers take responsibility for each child's learning, but it is not a responsibility to tell and initiate practice. It is a responsibility to reflect on the mathematical knowledge currently constructed by the child, to evaluate that knowledge in light of the content objective(s) and the desired understanding(s) that form the goals of the instruction, and to determine what questions or other problems should be posed to that child to facilitate and foster that child's developing mathematical knowledge.

For example, consider the conceptual understandings implied by two third graders' responses to a fraction task. The children were given a set of tangrams. They were asked to find one-fourth of the large triangle in the set. One third-grade girl moved the pieces about until she fitted two small triangles over half the area of the large triangle. She then said that the one small triangle was one-fourth of the large triangle. When asked why, she replied, "Because if you put these two [small triangles] on [the large triangle], then it ... that's one-half.... If you put these two on like this (fits the two small triangles over one-half of the large triangle), it would equal one-half. And, if you split one-half in half ... then it would equal one-fourth (holds up one small triangle)."

Contrast this to the explanation of a third-grade boy. He, too, moved the shapes about, identifying the square as being one-fourth of the large triangle because "it has four sides." Clearly these children's conceptualizations differ greatly. The challenge for a teacher is not to simply interpret the boy's response as incorrect. Rather, the teacher must consider what the boy feels a fraction entails, what aspect of his interpretation should first be confronted, and what task or problem may stimulate him to reconsider that aspect of his interpretation of fractions.

Meaningful learning takes place when the child is able to bridge the gap between existing mathematical and nonmathematical knowledge and newly encountered material, problems, or experiences. In a classroom, children will not all approach a problem in the same way, they will not all solve it in the same way, and they will not all take the same amount of time to reach their conclusions. But, if a problem is challenging without being frustrating to the child, if a problem causes the child to reflect on and to validate or modify current understandings, if a problem stimulates the child to connect ideas, then that problem is appropriate.

Over time, children need to progressively construct the increasingly abstract concepts and procedures that compose their mathematical understandings. Developmental maturity will determine the extent to which a child can integrate experiences as abstractions or connected structures (Case 1985).

Making sense

Research suggests that children learn mathematics when they connect their prior perspectives or insights, as constructed from their prior physical or mental experiences, to their next or new experiences. This connection is meaningful to the child because the child constructs it. While this sounds simple and straightforward as an expression of a theory, as the prior example indicated, it is not so simple when viewed in the reality of the classroom.

The connections constructed by the child may not be "correct" when viewed through the lens of an adult's perspective of mathematics, but they are meaningful to the child. The teacher must respect the understanding as constructed thus far by a child, yet, at the same time, the teacher feels a responsibility to uphold a standard for mathematics. Often times this leaves the teacher in a quandary. Teachers often state that they feel they have a responsibility to "set the child straight," to "fix it," to "tell the child what is supposed to be." Experience has shown that when teachers do tell children what to do, when teachers do "set the children straight," the teachers feel better; the teachers feel that they have done what they are supposed to do, they have met their responsibility.

However, experience also indicates that frequently the child knows no more than what the child knew before. The act of telling will not advance the child's understanding because what he or she was told probably does not make sense to the child. For example, consider the response offered by a second-grade child when presented the problem 25 + 37 = ___. First, she rewrote the problem vertically. After ten seconds she placed a small numeral "1" over the "2" in the number "25." After an additional forty seconds, she was questioned:

Teacher: Can I hear your thinking? You must be doing a lot of thinking. Whatcha thinking about?

Janine: How to carry the 1 because this is 12 (points to the 5 and the 7).

Teacher: That's 12, uh hmm. (46 seconds elapse.) Whatcha thinking?

Janine: Do I carry the 1 again?

Teacher: Hmmm?

Janine: Do I carry the 1 again?

Teacher: I don't know. Where did you get the 1 from in the first place?

Janine: From over here (points to the 5 and the 7).

Teacher: Uh huh. So, so now what do you do? ... What are you thinking about?

Janine: It's 11 over here (points to the 5 and the 7).

Teacher: It's ele ... (puzzled). Oh, I see. Then it's 11 now over here (points to the 5 and the 7) because you put 1 over there (points to the 1 over the 2). Okay. I heard, I heard another person one time use that word "carry" in a math problem. What does that mean? Who taught you about carrying?

Janine: My father did.

Teacher: Your father did. Okay. Thank you.

When children are told that their approach is incorrect and the "correct" method is demonstrated, a change has occurred. Now the child's belief that he or she can figure out how to solve a problem has been diminished. Now the child's perspective that it is his or her responsibility to complete the work involved in solving a problem is held up to self-ridicule. Now the child's perception that mathematics is about sense making is curtailed. It is critical to accept what a child has done as a meaningful, logical approach as viewed through the mathematical perspective currently constructed by the child.

However, learning will not be facilitated if teachers accept any response without further involvement simply because it is a reflection of that child's understanding. Asking the child questions, expecting the child to explain how he or she solved the problem and why he or she did it in that way, may give some insight as to the understandings constructed by the child. It may also prompt the child to reflect upon and possibly modify his or her thinking. If children persist in their interpretations, not recognizing the mathematical inconsistencies in their thinking, the teacher may confront them. "I don't understand. I want to understand what you did. Help me out here. Why did you...? How do you know that...? Tell me your thinking."

These questions may cause a child to face a misconception, to experience dissonance, but as a child attempts to address these questions, understanding is fostered. It may be that the thinking of the child then becomes modified. However, this is not always the case. Consider a fourth-grade child's interpretation of a problem addressing proportional thinking: 1/3 of 6 = 1/4 of ___.

> *Teacher:* It says that one-third of 6 ... is equal to one-fourth of some number. And I want to know, "What is the missing number?" I want you to think about what your answer would be if you took one-third of 6.... And now this problem says that there is some missing number so that if you took one-fourth of that missing number you would get the same answer as you get here. So what could that missing number be? ... And remember you can use this paper to draw on if you want.
>
> *Huong:* (Draws a circle cut into 6 sections. Shades in one-third of the sections. Repeats that process with a second circle, only her 6 pieces are more evenly divided. Draws a third circle separated into fourths. Shades one-fourth. Then she cuts each section in the third circle in half, creating 8 sections. Counts the 8 pieces.) ... Two-eighths.
>
> *Teacher:* Two-eighths? Okay, tell me what your thinking is.... Tell me what you're thinking.
>
> *Huong:* First I drew. First, I drew.... First I drew a circle with, I divided it into 6 parts. And I shaded one-third, so that equals 2 of the parts.
>
> *Teacher:* Okay, so one-third of 6 equals what?
>
> *Huong:* Two-o-o, two-sixths.
>
> *Teacher:* Okay.
>
> *Huong:* Two-sixths. And so that (first circle) equals then two-sixths. And, and I divided that (third circle) into.... Hmmm. (Draws another circle sepa-

rated into 8 parts. Shades in one-fourth.) ... Two-eighths.

Teacher: Okay.

Huong: So, I got two-eighths. And I just drew 6 ... 8 parts of, of a circle. So I just shaded in two- ... one- ... fourth. So I just shaded in one-fourth of the 8, that's the two-eighths.

Teacher: Okay, now how did you know to draw 8 parts?

Huong: I just ... I divi- ... I just drew lines across. (Draws another circle separated into 8 parts. Shades in 2 parts.)

Teacher: Okay.

Huong: I like the ... all the symmetries I can make. And I shaded, and I counted, and I got 8 parts. And I shaded in 2 parts. So that was one-fourth of it.

Teacher: Okay, so then here's my question. One-third of 6 ... is equal to one-fourth of what? (pointing to the written expression 1/3 of 6 = 1/4 of ___)

Huong: Two-eighths?

Teacher: Of two-eighths. Okay ... okay. Thank you.

This child's constructed knowledge has not changed, but it has become more explicit to the teacher. A constructivist approach then expects that the teacher will reflect on the kinds of problems or experiences that the child may need to stimulate further growth.

In order for a teacher's questioning to facilitate learning, it must be perceived by the child as an indication of the teacher's genuine interest in understanding what the child did and not as a hidden message that the child has made a mistake. This means a teacher should ask questions whenever a child offers a solution or an explanation to a problem. These questions must be asked when the child's solution is interpreted by the teacher as being "correct" and when the child's solution is interpreted by the teacher as being "incorrect." When this is done, children learn that mathematics is about making sense. Mathematics is about each child making sense of the problem and carrying through a meaningful approach, it is about the children making sense of each other's approaches, it is about the teacher trying to make sense of what the children have done.

Prior knowledge

All children come to school with prior mathematical knowledge. This may not be "academic" knowledge, but it is mathematical. In their study of preschool and kindergarten children in the Baltimore and Washington, D.C., area,

Ginsburg and Russell (1981) noted that mathematical thought developed in a "robust manner among lower- and middle-class children, black and white" (p. 56) and that prior studies that indicated "apparent incompetence in the young [or] the poor ... are more likely to be the results of the researcher's inadequacies than of the subject's. Careful attention to method and use of flexible, sensitive procedures often reveal unsuspected competence" (p. 53).

Project IMPACT's efforts to implement a constructivist approach to mathematics instruction in predominantly minority, urban schools has verified this perspective (Campbell and Rowan 1992, forthcoming). In this project meaningful mathematics instruction for all children is being accomplished through the context of child-centered problem solving. The approach used in this study is to address mathematical concepts in real-life problem situations that have meaning to the children. As children come to understand symbols, abstract computational representations can also become meaningful contexts for study, although young children may frequently draw pictures or make up word problems to provide meaning for those representations.

The maturation of mathematical understanding is best characterized as a process, not as an incident. It takes time; it does not happen all at once. Further, children do not develop mathematical maturity at the same rate. Therefore, in a primary or elementary classroom, children's prior knowledge will be manifested as differing levels of sophistication. To illustrate this, consider how two fourth-grade children solved this problem: Four children had 3 bags of M&M's. They decided to open all 3 bags of candy and share the M&M's fairly. There were 52 M&M candies in each bag. How many M&M candies did each child get?

Both children mentally determined that there were 156 candies to be shared by the four children. However, Marlena (see fig. 1) used symbolic notation to indicate her partial quotients and successive subtractions. She also distributed the 156 candies among the four children in only three steps, computing partial products at each step. Darrell (see fig. 2) used a tabular form to record his sequential partitionings and counted aloud as he distributed values so that he would know when he had distributed all 156 candies. Both children added their subtotals in order to determine that each child should receive 39 candies.

· ·

Figure 1. How Marlena Solved the Problem

Four children had 3 bags of M&M's. They decided to open all 3 bags of candy and share the M&M's fairly. There were 52 M&M candies in each bag. How many M&M candies did each child get?

· ·

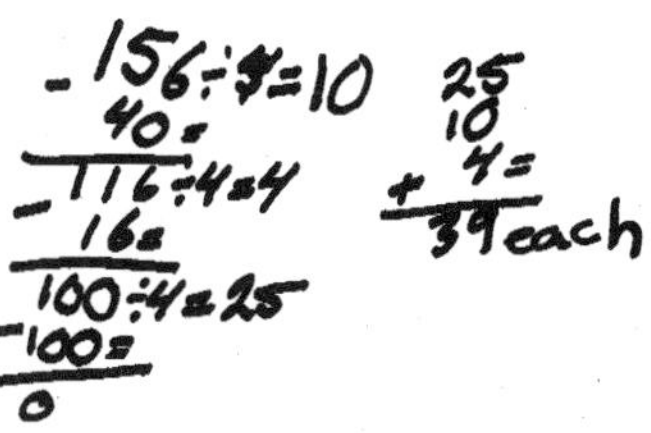

Figure 2. How Darrell Solved the Problem

Four children had 3 bags of M&M's. They decided to open all 3 bags of candy and share the M&M's fairly. There were 52 M&M candies in each bag. How many M&M candies did each child get?

It is critical to support each child's mathematical growth without limiting any child's opportunity to learn. Differentiating instruction within the heterogeneous classroom is a challenge to teachers, but it may be done within the culture of the classroom. The critical thesis is to support each child's efforts to develop, refine, broaden, and perfect his or her mathematical understandings over time as the child attempts to connect ideas, search for relationships, and apply mathematical concepts or procedures to real-life experiences.

SOCIAL INTERACTION CONSIDERATIONS

Intent of Instruction

Children need support in their efforts to construct meaning. This perspective is distinct from the policy of remediation that characterizes many elementary school programs. A remediation model focuses on the perceived deficiencies of a child, typically presenting a child with skill-oriented, direct instruction and practice as opposed to conceptually focused problem solving. The premise of the remediation model is that because a child is deemed to be wanting in some prerequisite skill or procedure, advancing mathematical investigations cannot occur until the deficit is made up.

Typically, application of the deficit model assures a continuing need for remediation. When a child is denied access to the continuing, advancing instruction of his or her peers while remediation occurs, then the child will necessarily remain behind those peers when the initial remediation is completed.

An alternative is to change the environment of the children's instruction, to maximize each child's efforts to think and to make sense of the mathematics. In particular, it is necessary to focus on the children's thinking and to direct attention to that thinking, as opposed to focusing on the expected algorithmic steps or procedural behavior and the correctness of the answer. The emphasis must be on children's learning, as opposed to emphasizing presentation of information and delivery of the curriculum. Children also need sufficient time to reflect and think, with the expectation that children will express and justify that thinking.

Further, children need to discuss their thinking and problem-solving strategies. When children express their ideas, teachers can reflect on their students' understandings and can ask additional questions to foster the students' thinking. Sometimes, children's explanations reveal a need for instruction to be adjusted in some way to support a child's learning. However, this modification should bolster the student's instruction, not limit it. For example, suppose a teacher decides that a group of children need more instruction to address misconceptions or not-yet-constructed ideas that the other children in the classroom have already developed. This instruction should be in addition to—not at the expense of—the instruction offered to the other children in the classroom.

To illustrate this further, consider the following. A fourth-grade teacher realizes that some of the children in her classroom are consistently erring if subtraction with regrouping is involved in the problem. She may then work with these children separately from the class's mathematics instruction period, targeting their construction of subtraction and place value while simultaneously addressing the topic of fractions with these children and the other children in the class during their regular mathematics instruction.

There is a culture in every classroom. It is critical that children find their classroom to be a place where ideas are accepted; suggestions are investigated; challenging, meaningful problems are solved; questions are asked to motivate children to construct understanding; questions are asked to build children's self-confidence and to celebrate children's ideas; and knowledge is "recreated, recycled and shared" (Ladson-Billings 1992, p. 114).

Shared Responsibility

While it is the student who ultimately constructs mathematical knowledge, this effort does not occur in a solitary setting nor is it totally autonomous. Learning can be catalyzed through interaction among children and between the children and the teacher. It is necessary for both the children and their teachers to take responsibility for learning meaningful mathematics.

When children work together in pairs or in small groups, they are expected to collaborate, to attempt to make sense of each other's efforts, and to explain or to question alternative approaches. This means that there is less emphasis on competition and more emphasis on each student taking responsibility for his or her learning and for sharing that knowledge with peers.

Following individual or cooperative efforts to solve a problem, whole-class sharing may encourage students to reflect on their thinking as they verbalize or demonstrate their strategy. However, to facilitate growth, a student's perspectives are not simply to be recited or accepted without comment. Learning may be enhanced if it is also presumed that each child who offers a strategy or approach is responsible for explaining or justifying that approach. Learning may be further stimulated in a classroom if students, as well as the teacher, question the ideas or thinking offered by another student, in an effort to understand the idea or strategy. In this way, the presumed validity of any solution can be examined and held to a standard of mathematical justification.

To learn mathematics is work. But, if it is the students, and not just the teachers, who do this work, then the students will gain mathematical power. As demonstrated in Project IMPACT classrooms, it is possible to promote responsibility for learning in mathematics. A distinctive feature in these classrooms is that the teachers try not to tell children how to solve mathematics problems nor do the teachers tell children if their solutions are correct or incorrect. Instead solutions are met with questions that seek to clarify and to verify the supporting mathematics. Further, no child is exempted from participating. No student is permitted to be a spectator. It is necessary for children to learn how to disagree without insult and how to ask questions without losing esteem. Each child's perspective must be valued, and the teacher must work with the children to establish mechanisms for classroom communication and behavior. But once these routines are established, children will take the risk to express and defend their opinion, children will learn to communicate mathematical ideas, and children will participate in mathematical inquiry. The key is to create a classroom community where every child is expected to respect each other's right to explain, to listen, and to ask questions.

Grouping Practices

It is essential that each classroom be a model of equity. This means more than promoting equity of opportunity. The standard must be that "educational experiences in the classroom support equal achievement and future participation in mathematics for all students regardless of their gender, ethnicity and race, or socioeconomic background" (Campbell and Langrall 1993, p. 110). All children must be expected to learn challenging mathematics.

Although few teachers or administrators would characterize their programs as limiting some children's access to mathematics instruction, the reality is that varied forms of ability grouping will exaggerate differences between students and lead to differentiated outcomes (Oakes 1985, 1990). The result is a limited opportunity to learn.

Children do not construct their mathematical understandings at the same rate or with the same level of sophistication. How, then, can the diversity of understandings in a classroom be addressed without resorting to ability grouping? The critical issue is to expand, not limit, each child's engagement. One approach is to use whole-class, heterogeneous instruction on some days, with small-group investigation for problem solving serving as the routine on other days. These small groups should frequently be heterogeneous, but they may also sometimes be homogeneous.

The critical distinction to support learning is that small-group composition is not always homogeneous. Whatever the definition of the small groups, following small-group interaction, whole-class sharing should still take place. To support learning, each child in the class, whether he or she solved a particular problem or not, must be permitted to hear and to question the solution strategies of the other groups. Each child must have the opportunity to learn from the approaches of the other children.

Consider the following example of homogeneous small-group instruction. The type of mathematics problem being investigated is identical across every group, only the size of the numbers differs. In a first-grade class two groups of children solve this problem: *Janine has 8 pennies. She wants to put 15 pennies in her bank. How many more pennies does she need?* Two other groups of children solve this problem: *Janine has 18 pennies. She wants to put 25 pennies in her bank. How many more pennies does she need?* A fifth group of children solves this problem: *Janine has 18 pennies. She wants to put 35 pennies in her bank. How many more pennies does she need?*

After the whole-class sharing takes place for each problem, the teacher may then ask the class how the problems were similar and different, as well as how the answers were similar and different. A discussion of the relationship between the numbers in the problems and the respective answers steers everyone in the class to consider notions of place value. The teacher may also ask how the suggested solution strategies were similar and different. In this way, all children are learning. The children not only have a chance to consider how a peer solved a problem that they themselves did not solve, they also have an opportunity to consider how numbers and solution strategies may generalize.

Another approach to support learning in diverse classrooms is to periodically use one or two homogeneous groups where the focus of the problems is different and more challenging or abstract from that in the other, possibly heterogeneous, groups, yet all groups are addressing the same mathematical topic. For example, a third-grade class may be exploring multiplication and division by creating multiple groups of the same size for a fixed amount of counters and then verifying the total value. After recording values to define the number of groups and the size of the groups as factors, the total number of counters may be modified in some predetermined fashion. Thus, the children are finding all possible factors of given numbers.

However, the teacher may decide that some of her students do not need further experience with this form of grouping and partitioning. These students may be told to work together to determine all of the numbers less than 50 that have an odd number of factors and then to characterize those numbers in some way.

Again, following completion of the tasks, whole-class sharing for all problems would take place. This not only validates the worth of each problem, it permits the teacher to challenge his or her children to explain the similarities and differences between the problems. This challenge, distinct from the actual problems, promotes the construction of further mathematical understandings, without overwhelming or limiting any child's engagement in the discussion.

FINAL COMMENTS

The discussion above makes it clear that teachers will approach children quite differently when teaching from a constructivist perspective. It has become very apparent, as we learn more about how children construct and learn mathematics and about the school environment needed to support such learning, that teacher education programs, both preservice and in-service, must take a greater responsibility for preparing teachers who are aware of such research and who are capable of using new research findings in developing effective mathematics programs.

The inclusion of new information into mathematics teaching has always been slow; however, encouragingly, the momentum is building for change as parents, administrators, and teachers demand higher levels of performance from students—performance that will ultimately depend on finding out more about how students learn mathematics and on using that information in our teaching.

. . . .

NOTE

The research reported in this material was supported in part by the National Science Foundation grant MDR-8954652. The opinions, conclusions, or recommendations expressed in these materials are those of the authors and do not necessarily reflect the views of the National Science Foundation.

. . . .

BIBLIOGRAPHY

Campbell, Patricia F., and Cynthia Langrall. "Making Equity a Reality in Classrooms." *Arithmetic Teacher* 41 (1993): 110–3.

Campbell, Patricia F., and Thomas E. Rowan. *Project IMPACT: Increasing the Mathematical Power of All Children and Teachers* (1991–92 annual report). College Park, Md.: University of Maryland at College Park, Center for Mathematics Education, 1992.

_____. *Project IMPACT: Increasing the Mathematical Power of All Children and Teachers* (final report). College Park, Md.: University of Maryland at College Park, Center for Mathematics Education, forthcoming.

Carpenter, Thomas P., and James M. Moser. "The Acquisition of Addition and Subtraction Concepts in Grades One through Three." *Journal for Research in Mathematics Education* 15 (1984): 179–202.

Case, Robbie. *Intellectual Development: Birth to Adulthood*. New York: Academic Press, 1985.

Ginsburg, Herbert P., and Robert L. Russell. *Social Class and Racial Influences on Early Mathematical Learning*. Monographs of the Society for Research in Child Development, 46 (6, Serial No. 193), 1981.

Hiebert, James, and Thomas P. Carpenter. "Learning and Teaching with Understanding." In *Handbook of Research on Mathematics Teaching and Learning*, edited by Douglas A. Grouws, pp. 65–97. New York: Macmillan, 1992.

Jackson, Phillip W. *Life in Classrooms*. New York: Teachers College Press, 1990.

Ladson-Billings, Gloria. "Culturally Relevant Teaching: The Key to Making Multicultural Education Work." In *Research and Multicultural Education: From the Margins to the Mainstream*, edited by Carl A. Grant, pp. 106–21. Bristol, Pa.: Falmer Press, 1992.

McKnight, Curtis, F. Joe Crosswhite, John Dossey, Edward Kifer, Jane Swafford, Kenneth Travers, and Thomas Cooney. *The Underachieving Curriculum: Assessing U.S. School Mathematics from an International Perspective*. Champaign, Ill.: Stipes Publishing Co., 1987.

National Council of Teachers of Mathematics. *Curriculum and Evaluation Standards for School Mathematics*. Reston, Va.: The Council, 1989.

National Research Council. *Everybody Counts: A Report to the Nation on the Future of Mathematics Education*. Washington, D.C.: National Academy of Sciences, 1989.

Oakes, Jeannie. *Keeping Track: How Schools Structure Inequality*. New Haven, Conn.: Yale University Press, 1985.

_____. *Multiplying Inequalities: The Effects of Race, Social Class, and Tracking on Opportunities to Learn Mathematics and Science*. Santa Monica, Calif.: Rand Corp., 1990.

Resnick, Lauren B. "A Developmental Theory of Number Understanding." In *The Development of Mathematical Thinking*, edited by Herbert P. Ginsburg, pp. 109–51. New York: Academic Press, 1983.

Riley, Mary S., James G. Greeno, and Joan I. Heller. "The Development of Children's Problem-Solving Ability in Arithmetic." In *The Development of Mathematical Thinking*, edited by Herbert P. Ginsburg, pp. 153–96. New York: Academic Press, 1983.

Rowan, Thomas E., and Barbara Bourne. *Thinking Like Mathematicians*. Portsmouth, N.H.: Heinemann, 1994.

What Secondary Mathematics Students *Can* Do

Raffaella Borasi

THE American public is by now quite used to hearing negative news about U.S. students' achievements in mathematics. Less than a decade ago, many people were shocked by international studies that showed how U.S. students scored among the lowest, among the nations tested, with respect to most measures of mathematics achievement (McKnight et al. 1987). The National Association for Educational Progress reports on the results of nationwide mathematics assessments (e.g., Dossey et al. 1988, 1993) have continued to show disappointing performances, especially in the area of problem solving, at all grade levels assessed (i.e., 4, 8, and 12).

These results may look especially disturbing when contrasted with recent recommendations for future school mathematics that point to the need of today's technological society for mathematical knowledge that goes beyond a mastery of the "basic skills" (e.g., NCTM 1989, 1991; National Research Council 1989). Consider, in particular, the ambitious goals articulated by the NCTM:

> [T]he K–12 standards articulate five general goals for all students:
>
> (1) that they learn to value mathematics,
>
> (2) that they become confident in their ability to do mathematics,
>
> (3) that they become mathematical problem solvers,
>
> (4) that they learn to communicate mathematically, and
>
> (5) that they learn to reason mathematically. (NCTM 1989, p. 5)

Raffaella Borasi is an associate professor in the Warner Graduate School of Education and Human Development of the University of Rochester. Her research has focused on the implications of an inquiry approach to mathematics education and on what happens when such an approach is implemented in a variety of instructional contexts.

Are these goals unrealistic for today's U.S. secondary school students? I firmly believe that this is not the case and hope to convince readers through the anecdotal evidence reported in this chapter.

With this goal in mind, I have selected four classroom experiences that I believe can illustrate how secondary school students can indeed achieve the goals set by the NCTM Standards—provided they are offered appropriate learning opportunities. While all of these vignettes will provide evidence of the students' ability to reason and communicate mathematically, each of them has been chosen to further highlight one of the following key instructional goals: becoming able to pose and solve novel mathematical problems, coming to value the potential of mathematics to "make sense" of real-life experiences, making connections between mathematics and other areas, and increasing confidence in one's mathematical ability.

The four vignettes have also been selected to represent a wide variety of student populations, instructional settings, teacher backgrounds, and content areas. They took place in an urban high school, a rural middle school, a suburban middle school, and a private school for severely learning disabled students. The classes were taught by teachers whose prior teaching experiences ranged from zero to ten years, and they dealt with different topics such as geometry, statistics, and measurement.

Though the focus of these vignettes will be mainly on the mathematical performance of the students portrayed in them, I have also tried to provide information on the instructional context and the kind of learning experiences the students were offered in these occasions. I will come back to highlight some of these instructional characteristics in the concluding section of this chapter, in the attempt to identify what curriculum choices and teaching practices may be necessary to reproduce the kind of outcomes illustrated in these episodes.

VIGNETTE 1. SECONDARY STUDENTS *CAN* LEARN TO POSE AND SOLVE NOVEL PROBLEMS CREATIVELY

This first vignette took place within a high school mathematics course taught by Judi Fonzi at the School Without Walls, an urban alternative high school in Rochester, N.Y. (See Borasi [in press] for a more detailed report on the instructional episode summarized in this vignette, as well as for other vignettes of this kind developed as part of the project "Using Errors as Springboards for Inquiry in Mathematics Instruction" [NSF award MDR-8651582].)

Typical of all Judi Fonzi's courses, as well as of most of the courses taught at the School Without Walls, an inquiry approach informed the teaching of the course. About twenty students, ranging from ninth to twelfth graders, were enrolled in this class.

The lesson reported here was developed within a unit on geometric constructions. In the previous few classes, the students had already succeeded, by using only ruler and compass, in constructing a triangle given two sides and the angle between them (SAS), a side and the two angles adjacent to it (ASA), and three sides (SSS), respectively, and in verifying that any triangle satisfying those conditions would be congruent to the first one.

Rather than reproducing constructions suggested by their textbook or by the teacher, in each case the students themselves had been expected to create their own construction and to justify their "method" to the rest of the class.

In the episode reported below, the students engaged in verifying by construction whether two angles and a side adjacent only to one of them are sufficient to determine a triangle uniquely. (This congruence criterion will be abbreviated as AAS hereafter.) This time, however, the task turned out to be more problematic than expected and ultimately resulted in two alternative procedures for constructing a triangle satisfying the given conditions.

The lesson began with the teacher going to the blackboard and asking the students to give her specific directions so that she could reproduce their suggested construction on the board. At first, one of the students (Linda) suggested drawing the two given angles (labeled 1 and 2) at the two opposite extremes of the given side (labeled 3), but then she corrected herself as she realized that this would not correctly represent the condition given (since one of the angles should *not* be adjacent to the given side):

> *Teacher:* Why are you changing your mind?

> *Linda:* Because for angle-angle-side you need the angle at the top.

A brief pause occurred while the teacher tried to follow Linda's idea literally, by reproducing angle 2 at the end of the extension of the other side of angle 1 (called "ray 1" hereafter), as shown in figure 1.

The students were quick to realize that this result was not acceptable, and some of them started to look for some ways to fix it. Sam first suggested "adjusting" line 3 so that the ray from angle 2 would meet it in the "right" spot. Most students in the class, however, were not willing to accept a "trial and error" procedure of this kind. Jane, in particular, was able to articulate precisely the key problem with this procedure:

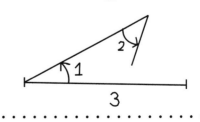

Figure 1. Construction Following Linda's Second Suggestion

> *Jane:* You've got to know the lengths of the sides before you add another angle to the [figure].

This precise articulation of the problem may have contributed to its solution, since at this point another student (Shea) was able to suggest a creative way to construct the desired triangle, through a procedure that made use of the first failed attempt:

> *Shea:* If you copy the angle that's right above [line] 3, that doesn't have a number ... *(she points to angle CC'A in figure 2).* If you copy that [angle] to the end of line 3, then just make the top of line 1 be the top of the other angle. Do you know what I mean?

The idea behind Shea's construction can be better understood by looking at her final product, reproduced in figure 2. (Note that the letters identifying specific points in fig. 2 have been added by the author so as to make the narrative that follows easier to understand.)

Figure 2. Shea's Construction

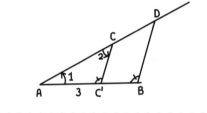

Shea's procedure is indeed a clever way to resolve the problem of knowing exactly how much ray 1 needs to be extended by taking explicit advantage of the information provided by a prior *incorrect* result, as she herself was able to realize and articulate:

> *Shea:* It works, but you can't do it unless you do it wrong first.

The rest of the class now seemed convinced that this construction would work and were able to repeat individually such a construction on another example upon the teacher's request. Some students were also able to explain why the procedure worked based on the observation that, since the angles *ACC'* and *ADB* (see fig. 2) are equal by construction, then the lines *CC'* and *DB* are parallel and, hence, the triangles *ACC'* and *ADB* are similar.

However, some students continued to argue among themselves about the appropriateness of the procedure itself. As the teacher overheard them, she decided to bring their controversy to the attention of the whole class:

> *Teacher:* Jim ... [you said to] Jane: "No, you don't have to do it wrong the first time to make it work." And Jane said, "So, how do you do it?"

Jim was thus invited to the board to prove his point. After some unsuccessful attempts, he had to give up. Another student (Todd), however, was ready to take his place at the board and show an alternative procedure to construct a triangle given AAS that would not require "doing it wrong first." His construction (see fig. 3) used the known property that the sum of the angles in a triangle is 180° to construct the third angle of the triangle and then used this angle to construct the triangle using the ASA procedure the class had developed in previous lessons.

Figure 3. Todd's Construction

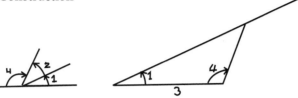

Todd was also able to clearly articulate the thinking behind this construction as follows:

> *Todd:* You see, the trouble was we couldn't get an angle at this end. And since all three angles of a triangle equal 180° *(he points to the picture he has just completed on the left)*, angle 1, angle 2 and you just ... the rest is that angle.
>
> *Pat:* Wow.
>
> *Shea:* That's really neat.

Both the teacher and the other students were quite impressed by the ingenuity and the novelty of this procedure, though a student questioned the validity of his procedure. This, in turn, initiated a discussion on why Todd's construction would work:

> *Student:* It does look like hers (Shea's). Hers just leaned a little more.
>
> *Mary:* Proof. Proof.
>
> *Teacher:* You've got to prove what, Mary? You've got to prove that the angle 2 that we were supposed to have really does end up there.
>
> *Student:* (interrupting) He did.

(As this conversation is taking place, Todd finishes his drawing and the class applauds.)

> *Teacher:* Who is the other person in this room who obviously thought about this picture and can explain it?
>
> *Jim:* He constructed angle 1 and then added angle 2 to it. Also three angles ... all three angles equal 180°. The portion left from angle 1 and angle 2 is ...

hmm, angle.... So then you go angle 1, line segment 3: [and then angle 4] goes from the end of line segment 3.

Teacher: Was angle 4 part of the given?

Student: No.

This observation allowed the teacher to explicitly point out that often in mathematics one has to use information and results that are not immediately given in the text of the problem. It also led to a final discussion of the difference between AAS and ASA, as well as to a more explicit appreciation of the cumulative nature of mathematical results, since previously proved results can be used in new proofs.

In this instructional episode, it is especially interesting to note how the students were not intimidated to approach a novel and challenging problem and were able to use their reasoning abilities as well as prior mathematical knowledge to propose and explore possible solutions (though they did not always turn out to be right). Communication was a crucial component of this lesson, since various students contributed ideas, evaluated other people's suggestions, and tried to convince each other of the appropriateness of a solution.

In this context, a convincing justification of one's results was also expected by everyone to be an integral part of the process, not just a requirement externally imposed by the teacher. It is also interesting to note that some students were not content with reaching an acceptable solution to the original problem but rather felt compelled and entitled to pose and explore yet another problem—how the construction could be achieved in an alternative way that did not require "doing it wrong" first.

VIGNETTE 2. SECONDARY STUDENTS *CAN* REGOGNIZE THE VALUE OF MATHEMATICS AND PUT THEIR MATHEMATICAL KNOWLEDGE TO USE IN REAL-LIFE SITUATIONS

In this vignette I will summarize the key elements of a three-week unit on "taking a census" that was designed and taught by Lisa Grasso Sandridge (at the time a first-year teacher) to her eighth-grade math class in a rural middle school about thirty miles from Rochester, N.Y. This experience developed as part of a larger project focusing on using "Reading to Learn Mathematics for Critical Thinking" (NSF award MDR-8850548). (See Borasi and Siegel [1994] for a more detailed description of this unit and Siegel et al. [in press] for vignettes of two other instructional episodes developed within this same project.)

The impetus for the unit was the concurrent 1990 U.S. census. The teacher saw this current event as an opportunity for her students to come to a better understanding of the role of mathematics in their everyday life while at the same time learning the rudiments of statistics in a meaningful context. With these goals in mind, she suggested that, while the U.S. census was being taken, the class could design and take a census of their own school. The students were given full responsibility for taking this school census, including choosing and formulating the questions to be asked, tabulating and analyzing the responses, and communicating the most significant results to the rest of the school.

Coming up with a "good" school census questionnaire already posed an interesting challenge, since the students had not only to come to a consensus about *what* questions to ask but also to carefully decide *how* to ask them, so as to be able to analyze their answers. As part of this process, the students carefully examined the formats used in the U.S. census long and short forms. The "school census questionnaire" they finally created (see fig. 4) was distributed and completed by all students in the school, during the morning homeroom period on the day after U.S. Census Day 1990. About five hundred students responded to the questionnaire.

. . . .

Figure 4. Students' Constructed School Census Questionnaire

CIRCLE THE CORRECT RESPONSE

1. Are you a   MALE   /   FEMALE

2. How many people live in your household

    1   2   3   4   5   6   7   8   9   10

3. Are your parents divorced   YES   /   NO

4. Does your family own a computer   YES   /   NO

5. How many people in your family watch "The Simpsons"

    1   2   3   4   5   6   7   8   9   10

6. Estimate your school average

    (A)   BELOW60,   B   60-70,   (C)   70-80,

    (D)   80-90,   (E)   90-100

7. How do you get to school

    WALK   BUS   DRIVEN   OTHER

8. Do you participate in any after school activities   YES   /   NO

    LIST 3 _____,   _____   _____

9. Do you participate in any after school sports   YES   /   NO

    LIST 3 _____,   _____   _____

10.   Do you presently have a   GIRLFRIEND   BOYFRIEND

. . . .

To directly involve everyone in the class in the analysis of these data, the teacher assigned each student a packet of completed census forms from a given homeroom. The student thus became the "enumerator" for that homeroom, responsible for tabulating the questionnaire responses of all the students in that homeroom, sharing those results with the rest of the class so that school-wide statistics could be created, and ultimately reporting the results back to the students in that homeroom.

The tabulation of the school census questionnaire was carried out in class and provided an opportunity for the teacher to introduce or review specific statistical concepts and techniques that could be used. For example, since the students found it difficult to tabulate the responses to the question "How many people live in your household?" (1, 2, ..., 10) because of the range of possible answers, the teacher introduced the notions of frequency and histograms, as well as the concepts of mean, median, and mode.

On another occasion, the realization that a number of students in the school had not completed the questionnaire led some students to raise questions about how representative their school census data were, a concern the class connected with the problem of "counting the homeless" in the national census they had previously read about in a newspaper article, and addressed through some readings from *How to Lie with Statistics* (Huff 1954).

At the end of each day, the teacher also entered into her computer the results of the students' tabulation to specific questions using a spreadsheet program—so as to update a table summarizing the key results of the school census by homeroom and grade level as well as school-wide—thus demonstrating the power of technology to represent and work with data.

A crucial aspect of the unit was the awareness that, in the end, each student was responsible for reporting back to the students in his or her assigned homeroom. To provide some support for this culminating activity, the teacher required each enumerator to prepare a poster summarizing what she or he thought were the most interesting results of the census taken in her or his homeroom, and to then use that poster as a guide for an oral presentation.

Though the teacher herself prepared a poster summarizing the results of the school-wide census as a model, each student was left to decide what questions to focus on in his or her poster, what statistics to use so as to effectively report specific results, what modes of representation to employ (summary statements, tables, graphs, and so on), and how to organize the information so that it would be both understandable and attractive. The teacher also used these posters as the main assessment tool for this unusual unit, along with a few homework assignments related to the analysis of the school census.

Indeed, the posters the students produced (see fig. 5 for some examples) showed that they had understood and could use appropriately a number of statistical concepts. Most students correctly employed frequency tables, bar graphs, and percents, while some even chose to use ratios or circle graphs in their posters. Many students also showed considerable creativity and artistic talent in putting together their posters—though the posters with more sophisticated mathematical information were not necessarily the more aesthetically appealing!

Figure 5. Sample of Students' Posters Reporting School Census Results from Their Homerooms

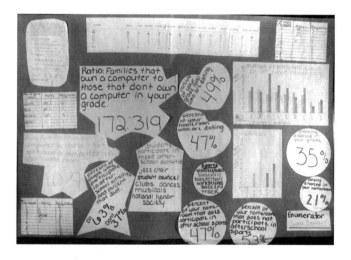

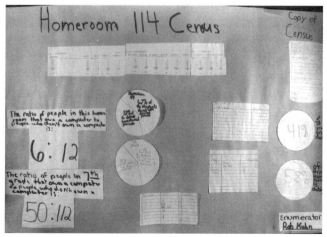

These posters, as well as the questions and insights the students had continued to generate throughout the analysis of the data, all confirmed that these students had been able to make sense of statistics in the context of their census project. This experience, along with the connections with the concurrent national census made at various points in the unit, also seemed to help the students better appreciate the role that mathematics could play in their own life.

VIGNETTE 3. SECONDARY STUDENTS *CAN* MAKE MULTIPLE MATHEMATICAL CONNECTIONS

The instructional experience reported here occurred at the end of a seven-week unit on tessellations developed collaboratively by Denise Anthony and Cynthia Callard, in a suburban middle school of the Rochester, N.Y., area, as part of the project "Supporting Middle School Learning Disabled Students in the Mainstream Mathematics Classroom" (NSF award TPE-9153812). (More information about the nature of this tessellation unit, its implementation in other instructional contexts, and various choices for appropriate assessment tools can be found in Fonzi and Rose [1994a, 1994b].) This tessellation unit was taught in all the six "regular eighth-grade math" classes in the school, including a "blended" classroom co-taught with a special education teacher. In this unit, the teachers used the topic of tessellation as a context in which the students could learn various geometric concepts and properties in a meaningful way, while at the same time exercising their ability to interpret mathematical definitions, make conjectures, and test them.

One of the culminating experiences in this unit consisted of creating a "poster" with an "Escher-like" tessellation. In previous classes, the math teacher had shown how M. C. Escher created even his most complex tessellations by generating a new tessellation shape as a result of appropriately "cutting and pasting" portions of a square or an equilateral triangle.

The class had then explored various ways of creating such "nibbles" themselves, making explicit connections with the kind of geometric transformations (translations, rotations, and so on) that made this possible. The art teacher had also been called in to provide more information about what made Escher's tessellations so aesthetically appealing.

With these background experiences and information, the students were given two weeks to prepare their own tessellation poster at home. The teachers had previously announced that this poster assignment would constitute a substantial part of their assessment in the unit (counting for about 20 percent of the final grade), and would be graded on the basis of multiple criteria, including creativity and artistic appeal.

Everyone was astounded at the quality and diversity of the posters the students turned in. (See the four posters reported in fig. 6 for a sample.) Indeed, in this assignment almost every student demonstrated a full understanding of tessellation and nibbling and, in many cases, artistic ability and creativity that surprised even the authors themselves! Some students who had not previously had much success in math classes were enabled to shine in this occasion; it is worth noting, for example, that one of the posters reproduced in figure 6 was created by a student with multiple learning disabilities.

Figure 6. Sample of Students' Tessellation Posters

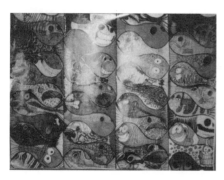

The positive experience of creating the posters in turn contributed to increasing these students' self-esteem as well as their peers' recognition of their potential contributions to the class as a learning community. Indeed, for several students this experience provided the opportunity for the first time to bring to their school mathematics work some of their strengths and interests—thus discovering new dimensions, meaning, and value for a discipline they might have previously felt was not for them. And this was further enhanced by the realization that this task was valued by the teacher to the extent of making it a part of their final evaluation in the unit.

VIGNETTE 4. ALL SECONDARY STUDENTS *CAN* INCREASE THEIR CONFIDENCE IN THEIR MATHEMATICAL ABILITY

This last experience took place in an eighth-grade mathematics class composed of eight students within a private school exclusively for severely learning disabled students. The teacher, Judi Fonzi, though an experienced mathematics teacher with over ten years of teaching experience at the time, was teaching for the first time in this instructional context as part of the previously mentioned project on supporting learning disabled students in mathematics. [A detailed narrative report of the series of over thirty experimental lessons taught by Fonzi in this class is provided in Borasi, Fonzi, and Smith [1994].)

The episode reported here developed over the course of two lessons within a unit on measurement. The students had initially been asked, as a homework assignment, to think about how measurement was "invented" and, more specifically, to write some examples of "what, why, and how" people might have measured in the past. As they shared the result of their work in class the next day, the whole class became intrigued with the specific problem of how cave dwellers might have been able to "measure" a cave in order to decide whether it was big enough for their family to live in for the winter.

Since they all agreed about the fact that cave dwellers would not have had yardsticks to measure with, and would probably not even know what our inches or feet are, the students were puzzled by *how* cave dwellers may have gone about measuring the cave. Some suggested that cave dwellers might have just "looked" at the cave and made a decision, though they were not able to justify how the cave dwellers might in fact have come to such a decision. Others suggested the use of sticks, or of the cave dwellers' own feet or arms, but did not go very far with these ideas and rather engaged in a digression about the size of cave dwellers (since some of them remembered reading that people have considerably increased in size over time).

There was also considerable debate about whether the cave dwellers would really *need* to measure a cave and why. John, a student who was usually rather quiet, participated more than usual in this discussion and finally suggested, "If it was the dead of winter, and you had no food, no nothing, do you think that they would really care ... how big this thing was? As long as they would all fit in it and be able to sit in it...." This observation helped the class come to the agreement that probably cave dwellers would want to find out, not the exact measure of the cave, but rather whether it would be big enough to hold everyone comfortably for a considerable period of time. This clarification finally enabled the students to proceed with the question of how such a decision could be reasonably and efficiently made. However, the class discussion that followed did not lead the students very far. The teacher then decided to assign the following homework for the next day:

> Write a story about a system cave dwellers might have developed for measuring to see if a cave is big enough.

When John shared his story in class the next day, he surprised everyone because of its ingenuity and clarity:

> The cave dweller would decide if there was enough room in a cave by first cutting down a tree a little longer than he is. Then he would hold it up and walk through the cave. If it touched the ceiling he would know that it was not high enough for his family. Then he would hold it lengthwise. If he did not touch the walls he would know he could sleep in the cave. Next he would flip it over end over end seven times, and if it didn't touch the end of the cave, he knew it would be a good cave. The stick was flipped seven times, once for each family member and two more times for extra room.

Upon the teacher's request, John was able to demonstrate his system very clearly to the rest of the class and even to address some of his classmates' criticisms regarding the accuracy of his system, as shown in the following excerpt from the class transcript:

> *Teacher:* John, could you show them? (*She hands him a yardstick.*) Could you pretend that this is a very short cave dweller and show us? (*to the rest of the class*) Do you know why I am asking John to do this?
>
> *Charlie:* Cuz it was his idea.
>
> *Teacher:* Yes, I am asking John to do it because it was his idea. (*Many of the students are asking to go on with the demonstration.*) But, hold on! The reason I am asking John to do this is because what he described was a pretty elaborated system and it is sometimes hard to draw pictures in your head about what he is talking about. So I am asking him to show us, so that we are sure we understand what he thinks. Go ahead, John, please.

(*John shows the class how one would walk through the cave holding the stick and flipping it over to measure various dimensions of the cave. All the other students are very excited and are all talking together.*)

> *Teacher:* Adam says it is not a feasible system, why?
>
> *Adam:* Because... what they would do is they would put the stick lengthwise, how would they know, they could have kids bigger than that. The kids could all grow bigger than their mothers.

Teacher: John, do you want to respond to that?...What would you do if a family member got bigger than the measuring stick?

John: Find another cave.

Teacher: Find another cave. And how would you do it? Would you take the same stick?

John: Get a bigger stick.

In light of John's story, and the thinking that went behind its creation and explanation, it may surprise the readers to know that this eighth grader had extremely severe computational problems (for example, he was not able to divide by 3 or to multiply 9 × 5 without a calculator) along with reading and writing disabilities (as illustrated by his previous homework assignment, reported in fig. 7). Indeed, his story had to be dictated to his mother, who at home used to scribe for him, and read by the teacher to the rest of the class.

. . . .

Figure 7. Sample of John's Handwriting in a Previous Math Assignment

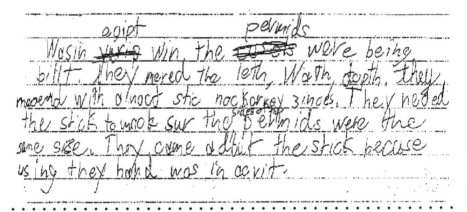

. . . .

This vignette illustrates how even severely learning disabled students like John can demonstrate their ability to reason and think mathematically once an appropriate learning task and the supports needed (such as using a calculator, having someone scribing, or using a word processor with spell-check capabilities) are provided.

There was an additional reason, however, why I chose to report it here. This episode constituted a turning point in John's experience in this classroom, as he himself began to realize that he, too, *could* do mathematics. With his newly gained confidence in his ability, he began to increasingly participate in the lessons and to be willing to engage in more and more challenging mathematics tasks.

WHAT INSTRUCTIONAL CONDITIONS CAN HELP SECONDARY STUDENTS MAKE FULL USE OF THEIR MATHEMATICAL POTENTIAL?

With all their obvious limitations, the vignettes developed in the previous pages have at least provided anecdotal evidence that secondary students *can* produce the kind of mathematical behaviors, attitudes, practices, and achievements called for by the NCTM Standards. (See Borasi [1992]; Borasi, Anthony, and Smith [1994]; and Healy [1993] for more in-depth and long-term illustrations.) In this concluding section of the chapter, I will identify some characteristics of the students' learning described in these vignettes as well as of the learning environments that helped produce these desirable results.

First, the episodes reported in this chapter differ most from what happens in typical mathematics classrooms because here the students were not expected to remember and apply taught facts and algorithms on routine mathematical exercises. Rather, in a variety of situations and topics, they actively engaged in "making sense" of problems that were novel to them, using their prior mathematical knowledge *along with* the common sense and thinking abilities that we all know adolescents are capable of (at least *outside* of school). While most of their peers, however, do not *believe* it is appropriate to bring those skills and life experiences to school mathematics (Borasi 1990; Schoenfeld 1992), these students did, and thus were able to be much more successful in mathematics.

Indeed, the youngsters portrayed in the previous four vignettes have all provided evidence of secondary students' ability to "construct" their own understanding of even complex mathematical concepts and, as a result, are better able to use those concepts appropriately in applications. These results are consistent with the theoretical considerations that form the basis of constructionist theories of learning (see, for example, Davis, Maher, and Nodding [1990] and von Glasersfeld [1991]) as well as with the results of empirical studies about how students solve problems and learn complex mathematics (see, for example, Resnick [1987] and Schoenfeld [1992]). As summarized in *Everybody Counts*:

> Educational research offers compelling evidence that students learn mathematics well only when they *construct* their own mathematical understanding. To understand what they learn, they must enact for themselves verbs that permeate the mathematics curriculum: "examine," "represent," "transform," "solve," "apply," "prove," "communicate." This happens most readily when students work in groups, engage in discussions, make presentations, and in other ways take charge of their learning. (National Research Council 1989, pp. 58–9)

It is also important, however, to realize that the mathematical behaviors illustrated in this chapter do not simply happen but rather require very intentional decisions and actions on the part of the teacher. Notice, for example, how the mathematical tasks the students engaged in were quite different from those typical of traditional math classes. Creating a novel geometric construction, analyzing data generated from a school census, creating a tessellation poster, or discussing issues in the history of mathematics are all much more complex and controversial activities than what mathematics students are used to.

The challenge and, occasionally, the controversy that these tasks presented—far from intimidating or frustrating the students—seemed to act as a powerful stimulus in all the episodes reported earlier (as suggested again by constructivist theories as well as supporters of an "inquiry" approach—see, for example, Borasi [in press]). At the same time, it is important to appreciate that this stimulus occurred because these tasks were quite open-ended and thus allowed for partial solutions as well as a variety of solution approaches, consequently enabling students with different abilities and strengths to find their own niche.

Also, the students rarely worked in isolation on these tasks and thus could build upon and benefit from the exchange of ideas and the different strengths contributed by each member of the class—a practice that has validated the increasing use of small groups and other forms of cooperative learning in all areas of instruction.

Another important element of these learning experiences is that the students always addressed mathematical topics and tasks within a meaningful context. This context was not necessarily a "real-life situation" (indeed only the census unit described in vignette 2 would fall into this category). However, in each case the "situation" the students dealt with (i.e., a complex geometric construction, the school census, the creation of the poster, the cave-dweller problem) was accessible and meaningful enough for them to be encouraged to draw upon the wealth of knowledge, experience, and thinking abilities that most students use in other domains of their lives but rarely think relevant to math classes.

The use of alternative forms of assessment—such as creating a tessellation poster, making a public report of their analysis of the census data, or even writing a story to explain how a mathematical system would work—was also crucial to making the students realize that the teacher really did not expect them to just regurgitate learned facts and algorithms. At the same time, these evaluative tasks enabled students with different strengths and learning styles to better demonstrate their learning as well as their potential mathematical ability.

It is also worth noting the multiple roles played by the teacher in the classrooms I have portrayed. Though all the teachers might have seemed more "invisible" in those situations—letting the students explore, make conjectures and objections, discuss, and come to their own conclusion—they were nevertheless indispensable to the success of those experiences. A lot of thought and planning went into the design of rich and stimulating tasks that would engage the students' interest and involve them in doing and learning some worthwhile mathematics.

The teacher also continued to play an important role as the activities developed in class—orchestrating discussions, facilitating the students' learning, or scaffolding when necessary. Most important, each teacher also devoted a lot of thinking and planning, even prior to the experience described here, to creating in her classroom a learning environment as well as a set of expectations and social norms that would be supportive of the new mathematical goals and behaviors they were trying to foster.

To conclude, my hope is that the experiences reported in this chapter will have provided reasons to be more optimistic about the future of mathematics education. U.S. secondary school students are indeed potentially more mathematically able than we give them credit for—though such a potential rarely shows in traditional math classes!

Our challenge—as teachers, administrators, parents, policymakers, and educational researchers—is to find effective ways to enable our students to recognize and capitalize upon their potential in school mathematics. This task is not an easy one, but it is by no means impossible, as shown by the encouraging experiences reported here, which are just a sample of the many I have witnessed lately in several secondary school math classrooms.

. . . .

NOTE

The instructional experiences reported in this chapter were developed within the following projects supported by grants from the National Science Foundation: "Using Errors as Springboards for Inquiry in Mathematics Instruction" (MDR-8651582); "Reading to Learn Mathematics for Critical Thinking" (MDR-8850548); and "Supporting Middle School Learning Disabled Students in the Mainstream Mathematics Classroom" (TPE-9153812). All the opinions expressed in this chapter are, however, the author's.

. . . .

BIBLIOGRAPHY

Borasi, Raffaella. *Learning Mathematics through Inquiry.* Portsmouth, N.H.: Heinemann, 1992.

_____. *Reconceiving Mathematics Education: A Focus on Errors.* Norwood, N.J.: Ablex, in press.

Borasi, Raffaella, Denise Anthony, and Catherine E. Smith. "Developing Area Formulas: An Opportunity for Inquiry within the Traditional Math Curriculum—In-Depth Story of a Classroom Experience." Preliminary report to the National Science Foundation for the project "Supporting Middle School Learning Disabled Students in the Mainstream Mathematics Classroom" (award TPE-9153812), 1994.

Borasi, Raffaella, Judith Fonzi, and Catherine E. Smith. "Detailed Narrative Reports of an Eight-Week Mathematics Experience with Severely Learning Disabled Students." Preliminary report to the National Science Foundation for the project "Supporting Middle School Learning Disabled Students in the Mainstream Mathematics Classroom" (award TPE-9153812), 1994.

Borasi, Raffaella, and Marjorie Siegel. "Reading, Writing and Mathematics: Rethinking the 'Basics' and Their Relationship." In *Selected Lectures from the Seventh International Congress on Mathematics Education,* edited by David F. Robitaille, D. H. Wheeler, and Carolyn Kieran, pp. 35–48. Quebec: Les Presses de l'Université Laval, 1994.

Davis, Robert B., Carolyn A. Maher, and Nel Noddings, eds. *Constructivist Views on the Teaching and Learning of Mathematics.* JRME Monograph no. 4. Reston, Va.: The Council, 1990.

Dossey, John A., Ina V. S. Mullis, and Chancey O. Jones. *Can Students Do Mathematical Problem Solving?* Washington, D.C.: U.S. Department of Education, 1993.

Dossey, John A., Ina V. S. Mullis, Mary M. Lindquist, and Donald L. Chambers. *The Mathematics Report Card: Are We Measuring Up?* Princeton, N.J.: Educational Testing Service, 1988.

Fonzi, Judith, and Barbara J. Rose. "Investigating Tessellations to Learn Geometry: Goals, Overview and Ideas for Planning the Unit." Preliminary report to the National Science Foundation for the project "Supporting Middle School Learning Disabled Students in the Mainstream Mathematics Classroom" (award TPE-9153812), 1994.

Fonzi, Judith, and Barbara J. Rose, eds. "Investigating Tessellations to Learn Geometry: Detailed Reports of Three Classroom Implementations." Preliminary report to the National Science Foundation for the project "Supporting Middle School Learning Disabled Students in the Mainstream Mathematics Classroom" (award TPE-9153812), 1994.

Healy, C. C. *Creating Miracles: A Story of Student Discovery.* Berkeley, Calif.: Key Curriculum Press, 1993.

Huff, Darrell. *How to Lie with Statistics.* New York: Norton, 1954.

McKnight, Curtis C., F. Joe Crosswhite, John A. Dossey, Edward Kifer, Jane O. Swafford, Kenneth J. Travers, and Thomas J. Cooney. *The Underachieving Curriculum.* Champaign, Ill.: Stipes, 1987.

National Council of Teachers of Mathematics. *Curriculum and Evaluation Standards for School Mathematics.* Reston, Va.: The Council, 1989.

_____. *Professional Standards for Teaching Mathematics.* Reston, Va.: The Council, 1991.

National Research Council, Mathematical Sciences Education Board. *Everybody Counts: A Report to the Nation on the Future of Mathematics Education.* Washington, D.C.: National Academy Press, 1989.

Resnick, Lauren B. *Education and Learning to Think.* Washington, D.C.: National Academy Press, 1978.

Schoenfeld, Alan H. "Learning to Think Mathematically: Problem Solving, Metacognition, and Sense Making in Mathematics." In *Handbook of Research on Mathematics Teaching and Learning,* edited by Douglas A. Grouws, pp. 334–70. New York: Macmillan, 1992.

Siegel, Marjorie, Raffaella Borasi, Judith Fonzi, Lisa Sandridge, and C. F. Smith. "The Unexplored Role of Reading in Mathematics Meaning-Making." In *Communication in Mathematics K–12,* 1996 Yearbook of the National Council of Teachers of Mathematics, edited by Portia C. Elliott. Reston, Va.: The Council, forthcoming.

von Glasersfeld, Ernst, ed. *Radical Constructivism in Mathematics Education.* Dordrecht, Netherlands: Kluwer, 1991.

Empowering All Students to Learn Mathematics

Gilbert J. Cuevas

THE road to reform in mathematics education leads to a future in which *all* students can be mathematically literate. The vision of *all* students having mathematical power has the potential for becoming a reality. But the path to change is not a smooth one, and the challenges to achieve the goals of reform are many. The potential obstacles include resistance to a change in the traditional expectations and attitudes we hold of students.

Obstacles may also appear in instructional practice, curriculum, student assessment, and teacher education. There are fears in the mathematics education community that the reforms will not reach all students. Of primary concern are student groups who have been historically underrepresented in mathematics. These groups include minorities and females.

The task at hand is to establish the necessary educational conditions to address the expressed concerns and overcome the existing and potential obstacles. To accomplish this will be a difficult task requiring extraordinary efforts. All actions must include bringing into the mathematics learning community all those students, with their diverse backgrounds and needs, who have been underrepresented in that community. Our involvement in reform cannot be merely the fine-tuning of existing endeavors. We must truly involve all students in the learning of mathematics. To accomplish this calls for the empowerment of students in the mathematics reform movement. Empowerment involves giving students the ability to control their own learning. This idea has a growing group of supporters in the mathematics education commu-

Gilbert Cuevas is a professor of mathematics education at the School of Education, University of Miami. He teaches undergraduate and graduate courses in mathematics education. He has published numerous articles on the teaching of mathematics to students with limited English proficiency.

nity. James Cummins (1989, p. 4) argues that reform in education is only possible when educators play an active role in empowering students. He elaborates:

> Students who are empowered by their interactions with educators experience a sense of control over their own lives and they develop the ability, confidence, and motivation to succeed academically. They participate competently in instruction as a result of having developed a confident cultural identity and appropriate strategies for accessing the information or resources they require in order to carry out academic tasks to which they are committed.

This chapter presents a brief overview of the need for the involvement of *all* students in the mathematics reform movement and the issues underlying the actions we may take to address this need. The discussion then addresses approaches educators can use to empower students in mathematics. The chapter concludes with the implications such approaches have for student assessment and teacher education.

THE NEED FOR STUDENT EMPOWERMENT

The makeup of the student population in our schools is very diverse. The Center for Educational Statistics reports that over 70 percent of the students attending the ten largest school districts are minority. In addition, 25 percent of the students in the United States are poor, and 20 percent live in single-parent homes. By the turn of the century between 30 percent and 40 percent of all the students enrolled in schools will be minority.

Against the background of these statistics, a disturbing picture of student achievement emerges. According to Biemiller, "the actual magnitude of observed educational skill diversity in our society is great—involving a range of more than four grades by grade 4" (1993, p. 8). Walter Secada reports that by age nine, white students outperform African Americans and Hispanics on national assessments of mathematics achievement. De La Rosa and Maw (1990) indicate that Hispanics have the highest dropout rate of any ethnic group in the United States. They also point out that only 51 percent of adult Hispanics have a high school diploma, compared to 63 percent for African Americans and 77 percent for whites. Our reform efforts must address student diversity concerns and the disparities in student achievement.

Reform in mathematics education also must overcome the obstacles that past instructional practices have promulgated: emphasis on memorization, drill, or what has been described as "ditto-sheet learning of isolated facts and skills." In a general sense, Cummins (1989, p. 58) has characterized the status of the education of minority students as follows:

...minority students are disempowered educationally in very much the same way that their communities are disempowered by interactions with societal institutions. The converse of this is that minority students will succeed educationally to the extent that the patterns of interactions in school reverse those that prevail in the society at large. In short, minority students are "empowered" or "disabled" as a direct result of their interactions with educators in the schools.

The reform movement in mathematics education calls for drastic changes in the way teachers interact with students. The NCTM *Professional Standards for Teaching Mathematics* (1991) presents a vision of mathematics teaching that emphasizes

- the selection of mathematical tasks that engage students' interests and intellect;
- the orchestration of classroom interactions (discourse) to promote the investigation and growth of mathematical ideas;
- the use of technology to pursue mathematical investigations;
- the structuring of small-group and whole-class work to help students understand the mathematics being studied and its applications;
- the promotion of connections to assist students in seeing the relevancy and applications of previous and developing knowledge.

The *Professional Teaching Standards* implies that if this view of teaching mathematics is to become real, certain obstacles must be overcome. These include the beliefs and attitudes that teachers and students have about learning mathematics as well as the assumptions that school administrators, parents, and society hold in general about teaching and learning, and specifically about mathematics.

The learning goals outlined in the NCTM's *Curriculum and Evaluation Standards for School Mathematics* (1989) also make the need for student empowerment more critical if we are to reach all students. It is only by giving students the tools to control their own learning that they will be able to

- apply their knowledge to solve problems within mathematics and other disciplines;
- use mathematical language to communicate ideas;
- reason and analyze mathematical problems;
- understand concepts and procedures;
- have a positive disposition toward mathematics;
- understand the nature of mathematics.

The development of "mathematical power" as communicated in the *Curriculum and Evaluation Standards* (1989) requires educators to examine a number of issues related to the extent to which all students will succeed in accomplishing these goals. Issues of equity, access, and excellence in mathematics education are at the core of mathematics education reform.

UNDERLYING ISSUES: WILL ALL STUDENTS REACH THE STANDARDS?

The Board of Directors of the NCTM affirmed the idea that the comprehensive mathematics education of *every child* is the most compelling goal of reform in mathematics education. By "every child" the Council meant

- students who have been denied access in any way to educational opportunities as well as those who have not;
- students who are African American, Hispanic, American Indian, and other minorities as well as those who are considered to be part of the majority;
- students who are female as well as those who are male;
- students who have not been successful in school as well as those who have been successful.

In addition, the NCTM (1986) has publicly stated through position statements its philosophy concerning the education of every child. The Council recommends that

> Mathematics educators must make an individual and organizational commitment to eliminate any psychological or institutional barriers to the study of mathematics. Innovative ways must be explored to convince both students and parents from underrepresented groups of the vital importance of mathematics courses in keeping both educational and career options open.

> Schools and districts whose enrollment in their most advanced mathematics classes does not reflect the overall demographic data for the school population should examine both their programs and their faculty for influences that might be leading to mathematics avoidance. Teachers at all educational levels should have the expectation that students from all segments of the population can be successful in mathematics. The teacher is in a key position to stimulate and encourage all students to continue the study of mathematics.

Mathematics has been viewed historically as the domain of upper- and middle-class male cultures. As a result, for females and minorities, mathematics has not played a significant part in their educational lives. Educational opportunity must accompany the need for *equitable* access to the content and experiences the students will have in the mathematics classroom. Both educational access and opportunity should go together with the belief that all students can excel. This relationship is described in the National Research

Council's *Everybody Counts: A Report to the Nation on the Future of Mathematics Education* (1989) as one which provides for all students "a full range of opportunities that can stimulate each person to tap fully his or her interests and capabilities," while stressing "that students achieve all that they are capable of accomplishing" (p. 29).

The notions of equity, equality, educational access, and excellence are at the center of our attempts to reach all students with mathematics. There are some of the opinion that the idea of equity has been dealt with lightly in many of the reform documents in mathematics education (e.g., NCTM's *Standards* documents, National Research Council's *Everybody Counts*). Secada (1989, p. 69) offers thought-provoking ideas on the relationship of educational equity to equality. He points out:

> Equity refers to our judgments about whether or not a given state of affairs is just. Equality in education often gets defined along some demographic characteristic: social class, race, gender, ethnicity, language background. Group differences are interpreted to demonstrate the existence of inequality. Equality, therefore, is defined implicitly as the absence of those differences.

It is important to reflect and keep in mind these ideas as we promote mathematical power for all students. We must remove obstacles, or unjust policies or curricula, that have prevented female and minority students from reaching their full potential in the learning of mathematics. In the context of the mathematics education reform movement, we must provide students with the skills and tools to empower each to become what Cummins (1989, p. 13) describes as

> ...an explorer of meaning, a critical and creative thinker who has contributions to make both in the classroom and in the world beyond.

Student empowerment in mathematics encompasses the notions of educational equity, equality, access, and excellence. It implies giving students control over their own lives as learners and the ability and confidence to make informed decisions about their own learning. Empowerment also involves helping students to use the mathematical knowledge they acquire outside of the classroom.

The teacher's role in the process of reaching all students is critical. Teachers must provide the appropriate environment so students can express, share, and broaden their learning experiences in mathematics. Principals, school board members, parents, and communities must also respond in kind to support teachers' efforts in making sure all students reach their full potential in mathematics. The next section discusses general principles to follow in our quest to empower all students in mathematics.

IMPLEMENTING CHANGE: OVERCOMING BARRIERS

A significant question for educators is, "How do we empower students to learn mathematics?" The processes involved are addressed explicitly and implicitly in both the *Curriculum and Evaluation Standards* (NCTM 1989) and the *Professional Teaching Standards* (NCTM 1991). This section provides a framework for examining these *Standards* documents from a student empowerment perspective. This framework includes three areas:

- Support of and respect for cultural diversity
- Nurture of critical mathematical literacy
- Development of meaningful classroom discourse

Cultural Diversity

Cultural issues lie at the heart of student empowerment. The extent to which students are empowered to learn mathematics depends on the types of interactions that take place in the schools and the mathematics classroom. Institutionally, schools have the choice of addressing the students' cultures along what Cummins (1989, p. 60) calls an additive-subtractive dimension. According to him,

> educators who see their roles as adding a ... cultural affiliation to students' repertoire are likely to empower students more than those who see their role as replacing or assimilating them to the dominant culture.

We must follow an additive approach to the integration of culture in the classroom. This contrasts significantly with the opposing or subtractive perspective of fitting all students into a homogeneous cultural mold.

Diversity is also reflected in student differences in learning modes and their knowledge and views of the world. Culture encompasses these features among others. Our efforts to empower students to learn mathematics must address these aspects of culture.

The NCTM *Professional Teaching Standards* (1991) calls for "making connections with other school subjects and with everyday society." This idea of connections can be extended to cultural diversity strategies. How can we connect the mathematics we teach with the cultural makeup of our students? To make instruction culturally sensitive to our students, Donovan (1990) offers these general principles:

- Recognize that all students regardless of cultural background, socioeconomic class, gender, and race are capable of learning and appreciating mathematics.
- Promote classroom discourse in which students play an important role in the interactions that take place. The students must feel their contri-

butions are valued and respected.
- Emphasize student responsibility in learning.
- Give students opportunities to use their own cultural backgrounds as a basis for the meaningful learning of mathematics. This can be accomplished by using culture as a context for mathematical explorations and problem solving.

In addition, there are four areas we can address to make specific connections with the cultural diversity of our students. First, mathematics contains rich historical roots. Unfortunately, only Western European contributions to the development of mathematics are included in most texts. Students need to know that other civilizations—Asian, Mexican, Egyptian—have made significant developments in mathematics.

Second, mathematics can be a tool or vehicle to study other cultures and countries. For example, students can learn about other people through the use of learning activities that focus on specific characteristics of their countries.

Third, students must be aware that others within their cultural groups use or develop mathematics as part of their livelihoods. Programs already exist in many school districts that bring in professionals from a variety of ethnic and cultural backgrounds to work with students. Students need meaningful role models so mathematics can become more real.

Fourth, linkages need to be established between the school and the home. Cummins has argued that "minority students will be empowered in the school context to the extent that the community themselves are empowered through their interactions with the school" (p. 62). Besides serving as cultural resources, parents need to become partners with the schools in the education of their children.

Lily Wong-Fillmore (1983) indicates that most parents of children from underrepresented groups in mathematics have high aspirations for their children. Unfortunately, educators at times have not involved parents effectively in the education of their children. In addition, parents have been prevented from participating in school matters for reasons beyond the school's control (e.g., work schedule conflicts with school events, lack of child care). The literature has also documented the positive changes that take place in children's academic progress when parents become active participants in educational efforts.

Critical Mathematical Literacy

The integration of cultural issues and topics in mathematics instruction must take place in the context of a teaching environment that emphasizes the students as active generators of their own knowledge. The *Curriculum and Evaluation Standards* (1989) focuses on approaches that promote such active participation.

To empower students, teachers must provide opportunities for students to express, share, and apply mathematical ideas. This notion is embedded in the curriculum and instructional recommendations of the *Standards*. To accomplish this, teachers first must become empowered themselves to implement such instructional reforms.

Teachers must have the necessary resources, knowledge, and support to create the required classroom environment. Schifter and Fosnot (1993, p. 37) illustrate this reform-conducive environment in their report of a teacher's perceptions about her interactions with students:

> I see myself as a facilitator.... Children should be provided many opportunities to develop social skills such as cooperating, helping, negotiating, and talking with other persons involved in solving problems.... The quality of learning that takes place from self-directed problem-solving and experimentation is important.... If the environment that we set up for learning is meaningful to [the students], they will be motivated enough so that they will try to create meaning out of whatever it is that we are presenting to them.

This process of creating appropriate classroom settings and giving all students opportunities to construct understanding is related to the development of critical literacy in mathematics. Being mathematically literate in the 1990s means having the power to use mathematics for our own purposes as well as those that the institutions of our society require of us.

As students develop this literacy, they must be encouraged to explore and analyze critically the mathematical content they are learning. This aspect of mathematical literacy is addressed in the *Professional Teaching Standards* (1991) in the context of the development of problem solving, communication, and reasoning skills.

The development of mathematical critical literacy skills also includes

- the development and application of strategies to solve a wide variety of problems;
- the verification and interpretation of problem-solving results;
- the reflection, representation, discussion, and clarification of thinking about mathematical ideas and situations;
- the drawing of logical conclusions;
- the use of models, known facts, properties, and relationships to explain thinking about mathematics;
- the justification of answers and solution processes;
- the use of patterns and relationships to analyze mathematical situations.

Classroom Discourse

Reaching all students with mathematics also involves empowering them to communicate proficiently about the content they are learning. Meaningful communication must occur within the context of everything that takes place in the mathematics classroom. This context has been referred to as *classroom discourse*. *Professional Standards for Teaching Mathematics* (1991, p. 34) describes classroom discourse as

> the ways of representing, thinking, talking, agreeing, and disagreeing.... Discourse is both the way ideas are exchanged and what the ideas entail.... The discourse is shaped by the tasks in which students engage and the nature of the learning environment.

The challenge for teachers is to develop classroom discourse in a group of students who are diverse in mathematical achievement, communication skills, and learning experiences. To meet this challenge, teachers and students have very definite roles to assume. *Professional Teaching Standards* offers a number of guidelines promoting classroom discourse, and they can be adapted to address the diversity of student backgrounds. Teachers should do the following:

- Pose questions and tasks in ways that challenge and engage students' thinking. Avoid questions that require single-word responses, or follow such questions with requests for students to explain their answers or to conjecture about other responses.
- Listen carefully to students' questions and answers, and ask for clarification when the questions and responses are not clear or relevant. Develop the habit of rephrasing the students' questions. At times, the students need to hear their own questions posed and decide for themselves if the questions are not clear or relevant. Also, provide a "buddy" for those students who may not be sufficiently fluent in English. This buddy or helper could assist in the translation of questions.
- Monitor students' participation in class activities. Be sure all students in the classroom have equal opportunities to participate in classroom discourse.
- Facilitate student involvement in class discussions and interactions. You may wish to engineer working groups in such a way that there is diversity of skills and backgrounds in each of them. Communicate to the students your expectations about classroom participation in a non-threatening way.
- Provide assistance when students are not able to participate in discourse or cannot express themselves appropriately. This might involve a classmate who can act as a helper for certain students as well as an instructional aide who can communicate with limited-English-proficient

students in their native language.
- Decide when to provide information, when to clarify an issue, when to give special assistance, when and how to model appropriate learning behaviors, and when to let students struggle with difficulties.

All students also have a role in the development of classroom discourse in mathematics. They must

- listen to, respond to, and question the teacher and one another respectfully;
- understand the elements of a given task or problem, and request help whenever the objectives of a task are not clear;
- use the resources provided by the teacher to solve problems and to communicate to others the results of activities;
- present solutions and convincing arguments for particular approaches used in problems.

Mathematical discourse and its accompanying classroom climate are not achieved overnight. It is through the joint efforts of teachers and students working on meaningful mathematical tasks that a community of learners can emerge.

IMPLICATIONS FOR MATHEMATICS EDUCATION IN THE TWENTY-FIRST CENTURY

The reform movement in mathematics education, together with the need for student empowerment in learning, affects the way we assess students, the uses of technology in the classroom, and the processes used in teacher education.

Student Assessment

The reliability and validity of traditional achievement tests have been questioned from legal and research perspectives. Courts have ruled that many of these assessment instruments lack curricular validity. In other words, the tests include content not addressed in the school's program of studies. Failure rates in standardized tests among African American and Hispanic students have been higher than those for white students. The courts have stated that if a test addresses material the students had not covered in class, it is unfair and violates the equity and due process rights of the students.

The potential for race, gender, and ethnic group discrimination exists in assessment systems designed to address reforms in mathematics education. The Mathematical Sciences Education Board (MSEB) in *Measuring What*

Counts: A Conceptual Guide for Mathematics Assessment stresses that changes in assessment must be sensitive to the diversity of students who will be the object of assessment practices.

The MSEB proposed that "assessment should support every student's opportunity to learn important mathematics" as an "equity principle" to be followed in all we do to measure student learning. The challenge is how to make this possible when addressing a student population that is increasingly diverse. The MSEB (National Research Council 1993, p. 93) strongly recommends that

> just as good instruction accommodates differences in the way learners construct knowledge, good assessments accommodate differences in the ways students think about and display mathematical understanding.

Therefore, to empower students to demonstrate the mathematics they have learned, multiple opportunities and a variety of assessment strategies must be offered. In addition, assessment practice must consider the students' backgrounds and experiences. In *Assessment Standards for School Mathematics*, the NCTM (1993, p. 41) elaborates on this issue by emphasizing that

> students' knowledge and ways of thinking and learning about mathematics are formed through a complex integration of their background with their experiences in school. Judgments about students' mathematics learning must consider the ways in which their unique qualities influence their ways of knowing mathematics and of communicating that knowledge.

Recommendations for making assessment practices more equitable have also been made by measurement experts. Linn (as reported by Rothman [1994]) suggests giving students a choice in the selection of assessment tasks. This includes not only the nature of the task (e.g., constructing a table, interpreting and summarizing data) but also the format and even the language used in the presentation of the task. For example, bilingual students may wish to consider a choice of language in responding to assessment tasks.

One basic message that echoes throughout the education community is that reform in assessment practices will be a long and difficult process that must involve all stakeholders in the education community.

Use of Technology

The NCTM *Professional Teaching Standards* (1991) stress the availability of computers in every mathematics classroom as well as the need for students to have access to the technology for both individual and group work. The document also emphasizes the role of the computer as a tool for processing information, performing calculations, and problem solving. The NCTM also recommends the integration of calculators into the school mathematics program at all grade levels in classwork, homework, and evaluation. Underlying the use of technology in mathematics education—namely, computers and calculators—is the assumption that *all* students will have access to the technologies and that *all* students will benefit from their use.

Inequalities based on economic resources have been reported concerning the use of technology in schools. Dublin et al. (1994) comment that in schools in affluent districts, students more frequently use computers for creative explorations and higher-order-thinking tasks. These authors also indicate an emphasis on drill-and-practice routines in mathematics, reading, and writing for students in less affluent schools. Providing equitable technological resources—computers and calculators—is one of the many challenges facing reformers in mathematics education. Many are concerned that while technology raised expectations for greater equity and quality in education, in reality, its implementation may have widened the learning gap between the haves and the have-nots.

Issues of equity and technology revolve around the distribution of technological resources and the use of such resources once they find their way into the mathematics classroom. The imbalance between rich and poor districts when it comes to the availability of computers and calculators is evidence of inequity. While funding programs have contributed somewhat to restoring balance in the availability of technology, the efforts have not exactly been far reaching.

The roles technology plays in learning and how teachers use technology with students are concerns when addressing equity issues. Are computers and calculators used as tools for the enhancement of learning, or are they assigned a mindless role as a vehicle for drill and practice? Some educators report that in many districts, given the emphasis to improve standardized test scores, computers are relegated to the role of "practicing to the test." Also, because of the fear of test achievement failure, calculators are forbidden in many mathematics classes.

Do female and minority students get equal access and exposure to computer and calculator activities in mathematics? Some suggest that the use of these technologies is more frequently found with males and the "more mathematically able" students. Critics have pointed out that students from groups traditionally underrepresented in mathematics are not involved in computer activities in the classroom. Others are concerned about the manner in which females and minorities are portrayed in software (which is developed mostly by white male computer programmers).

The issues will be difficult to resolve. It takes continuous, concerted efforts by mathematics educators and policymakers to achieve equity in the instructional uses of technology. Classroom teachers can contribute to this goal by addressing a number of ideas and strategies derived from the *Curriculum and Evaluation Standards* (NCTM 1989):

1. Begin the implementation of computers and calculators in the mathematics classroom with the belief that technology is a tool whose main purpose is to enhance the learning of all students, regardless of gender, race, achievement level, or handicap.
2. The physical environment of the classroom must be conducive to the use of a variety of instructional strategies, many of which must include computers and calculators.
3. Support a learning climate in which students are free to experiment with computers and calculators, and where mistakes are viewed as learning opportunities.
4. Promote administrative and parental support for the appropriate availability of computers and calculators. Encourage other teachers to join in the efforts to obtain the appropriate technological resources needed by all students in mathematics.

Teacher Education

Reform in the teaching of mathematics, with its underlying need to empower students' learning, must accompany changes in teacher development and education. If one of the goals of pedagogical practice is to reach all students with mathematics, then teacher education must focus on empowering teachers to carry out the mandates of reform.

Cummins (1989) cites the following general principles for teacher education if the goals of teacher and student empowerment are to be achieved. He states that in addition to teacher content knowledge in a given discipline, programs must prepare teachers to

- guide and facilitate students' learning rather than control it;
- encourage students to work collaboratively in a learning context;
- assist students to establish genuine dialogue through a variety of modalities (e.g., written, oral, through concrete models);
- facilitate the development of students' higher-level cognitive skills;
- plan, design, and implement tasks that generate intrinsic rather than extrinsic motivation.

More specifically to mathematics, the professional development of teachers must have a number of characteristics. Programs at both the preservice and in-service levels must *model good mathematics teaching with diverse student groups*. Teachers must be equipped to address the variety of learning needs and styles of the student populations they face in their classrooms. Teachers must be familiar with mathematical pedagogy and how content can be most effectively presented to students having a variety of mathematical, linguistic, and cultural backgrounds.

Programs must give teachers a *thorough knowledge of mathematics and school mathematics* as well as the *knowledge concerning diverse students as learners of mathematics.*

Overall, the traditional ways in which minority students have been and continue to be taught have resulted in their experiences being systematically excluded from the curriculum and the classroom. Teacher development programs must have as a guiding theme the empowerment of all students to learn mathematics.

In closing, student empowerment is a means to reach all students to achieve their highest potential in mathematics. Empowering students is not the sole responsibility of classroom teachers or students but also of policymakers, teacher education institutions, and community and parents. Each of them must act in concert to promote and bring into reality the vision of students who exemplify the knowledge of, and appreciation for, mathematics embodied in the NCTM *Standards* documents.

Initiatives are needed to make sure underrepresented student groups can take their rightful place in the mathematics reform movement. First, *we must make the education of these students a priority.* We must not treat these issues as a peripheral concern to be dealt with by a "subgroup of educators." The issue of equity and student empowerment in mathematics must be the concern of all. It must cut across curriculum programs, staff development, and research efforts.

Second, educators must recognize that *underrepresented student groups have a diversity of characteristics and needs.* There is no single strategy that can address all these needs. It is best to use multiple approaches in the process of student empowerment in mathematics.

Finally, *teachers', administrators', and policymakers' beliefs as well as community members' priorities need to be compatible with the envisioned changes and the need to include all students in the reform.* Teachers must have high expectations for all students. They must hold to the view that all students can learn and it is their right to learn. Educators must critically examine their views of what mathematics is and their knowledge and perceptions of the processes for learning mathematics.

· · · · · · · · · · · · · · · ·

BIBLIOGRAPHY

Biemiller, Andrew. "Lake Wobegon Revisited: On Diversity and Education." *Educational Researcher* 22 (9) (1993): 7–12.

Center for Educational Statistics. *Condition of Education.* Washington, D.C.: Government Printing Office, 1987.

Cuevas, Gilbert, and Mark Driscoll. *Reaching All Students with Mathematics.* Reston, Va.: The Council, 1993.

Cummins, James. *Empowering Minority Students.* Sacramento, Calif.: California Association for Bilingual Education, 1989.

De La Rosa, D. A., and C. T. Maw. *Hispanic Education: A Statistical Portrait.* Washington, D.C.: National Council of La Raza. Columbus, Ohio: ERIC/SMEAC Clearinghouse (ED 325 562), 1990.

Donovan, Brian F. "Cultural Power and the Defining of School Mathematics: A Case Study." In *Teaching and Learning Mathematics in the 1990s,* 1990 Yearbook of the National Council of Teachers of Mathematics, edited by Thomas J. Cooney, pp. 166–75. Reston, Va.: The Council, 1990.

Dublin, Peter, Harvey Pressman, Eileen Barnett, Ann D. Corcoran, and Evelyn J. Woldman. *Integrating Computers in Your Classroom: Elementary Education.* New York: HarperCollins College Publishers, 1994.

Kennedy, M. M., R. K. Jung, and M. E. Orland. *Poverty, Achievement, and the Distribution of Compensatory Education Services.* Washington, D.C.: Government Printing Office, 1986.

McLeod, A. M. "Critical Literacy: Taking Control of Our Own Lives." *Language Arts* 63 (1986): 37–50.

National Council of Teachers of Mathematics. *Assessment Standards for School Mathematics,* Working draft. Reston, Va.: The Council, 1993.

———. *Curriculum and Evaluation Standards for School Mathematics.* Reston, Va.: The Council, 1989.

———. *The Mathematics Education of Underrepresented Groups.* Position statement. Reston, Va.: The Council, 1986.

———. *Professional Standards for Teaching Mathematics.* Reston, Va.: The Council, 1991.

National Research Council, Mathematical Sciences Education Board. *Everybody Counts: A Report to the Nation on the Future of Mathematics Education.* Washington, D.C.: National Academy Press, 1989.

———. *Measuring What Counts: A Conceptual Guide for Mathematics Assessment.* Washington, D.C.: National Academy Press, 1993.

Rothman, Robert. "Assessment Questions: Equity Answers." Evaluation comment. *Newsletter of the Center for the Study of Evaluation and the National Center for Research on Evaluation, Standards, and Student Testing* (Winter 1994): 1–12.

Schifter, Deborah, and Catherine T. Fosnot. *Reconstructing Mathematics Education.* New York: Teachers College Press, 1993.

Secada, Walter G. "Educational Equity versus Equality of Education: An Alternative Conception." In *Equity in Education,* edited by Walter G. Secada, pp. 68–88. New York: Falmer Press, 1989.

_____. "Race, Ethnicity, Social Class, Language, and Achievement in Mathematics." In *Handbook of Research on Mathematics Teaching and Learning,* edited by Douglas A. Grouws, pp. 623–60. New York: Macmillan Publishing Co., 1992.

Wong-Fillmore, Lily. "The Language Learner as an Individual: Implications of Research on Individual Differences for the ESL Teacher." In *On TESOL (Teachers of English to Speakers of Other Languages) '82: Pacific Perspectives on Language Learning and Teaching,* edited by M. A. Clarke and J. P. Handscombe, pp. 34–55. Washington, D.C.: TESOL, 1983.

Opportunity to Learn: Can Standards-Based Reform Be Equity-Based Reform?

Jeannie Oakes

T HE move toward national standards and performance-based accountability got a big boost last year when both the House and the Senate passed President Clinton's Goals 2000 Act. With federal dollars and the guidance of a national standards council, states are embarking on efforts to develop and submit for national certification new and higher standards for what U.S. students should know.

Despite increasing political enthusiasm for performance-based standards reforms as a strategy to increase the quality of education, critical questions have been raised by many analysts, including Undersecretary of Education Marshall Smith, about the distribution of these expected effects.

- How will low-income, minority students now trapped in schools with unchallenging curricula and inadequate learning conditions fare under such reforms?
- Will these students be better or worse off with standards-based schooling than under the current system?
- How might the distribution of resources and opportunity be affected?
- How might such a system treat linguistic and cultural differences?
- Can protections for low-income and minority students be built into such a system?

Jeannie Oakes is professor of education and assistant dean for teacher education programs in the Graduate School of Education and Information Studies at the University of California at Los Angeles and a consultant at the RAND Corporation. Her research and writing focus on how state and local policies affect the opportunities of minority and disadvantaged students.

Many standards advocates argue that if national content standards are accompanied by high-stakes performance assessments and accountability mechanisms (a strategy that will be referred to hereafter as *performance-based standards*), they will actually make schools more responsive to the special learning needs of disadvantaged and minority students—that schools will "rise to the occasion." This optimism rests partly on plans for linking incentives for teachers and administrators to students' scores.

However, advocates also anticipate that the process of specifying high standards in each of the major content areas and developing consensual legitimacy and authority for these standards will, in itself, raise teacher expectations and increase student motivation and effort. Lauren Resnick (1993), for example, has argued that such reforms will generate unprecedented energy and commitment to students' "right to achieve" and create conditions wherein schools feel morally obligated to serve all students well.

Toward this end, states and school systems participating in the New Standards Project, for example, agree to a "social compact" in which they "pledge to do everything in our power to ensure all students a fair shot at reaching the new performance standards." They also promise that students who attend schools without sufficient opportunities will not be subjected to high-stakes decisions based on their performance.

Opponents of the standards movement counter with a variety of arguments ranging from traditionalists' concerns about eroding local curriculum control to critical multiculturalists' fear of further reifying the white, Western cultural canon. The loudest objections have been raised by advocates for low-income minority children who fear that the hoped-for motivational power of high performance-based standards will pale in comparison to the force of historical isolation and inequalities in schools serving low-income minority students.

These advocates argue that the cultural capital that white and wealthy students bring from home privileges them in the competition to do well on high-stakes performance assessments and that the technology for developing culturally fair assessments of student performance in the proposed "thinking" curricula is so underdeveloped that low-income minority students cannot possibly be either judged or treated fairly in such a system for many years to come.

For support, some point to the proliferation of standardized testing and credentialing over the past twenty-five years and the lack of evidence that these de facto national standards have led to improved schooling, greater student effort, or higher achievement for disadvantaged students. They argue that what low-income minority students and their schools need most is a solid educational infrastructure that supports educators' moral commitment to students' achievement. (See Clune [1993] and Darling-Hammond [1991] for good discussions of these and other objections to reforms driven by standards assessment and performance assessment.)

Fundamental disagreements about the nature of equality and fairness lie behind these specific arguments. As political scientist Lorraine McDonnell (1993) has observed, those who hold an ideology supporting performance-based standards and accountability as a powerful engine that can drive both education quality and equity are, for the most part, oriented toward *procedural* fairness.

This traditional "due process" view makes provisions for equity and disadvantaged schools primarily a matter of technical adjustments—for example, insuring that assessments and their administration are culturally fair and that factors beyond the ken of schooling (e.g., language-minority status or poverty) don't confound judgments about how well various schools are doing with their students. It's not surprising, then, that this side gives primacy to performance-based standards and promises to accommodate equity and opportunities.

The other side, primarily researchers and practitioners whose principal work and orientation is focused on those of the nations' schools and children who are not white and advantaged, emphasizes *substantive* fairness—whether or not all students have access to a fair share of the educational pie. Following a less traditional definition of equity, they tend to identify a fair distribution of resources and opportunities as the engine that will best drive educational improvement and equity. It's not surprising, then, that they tend to view performance-based standards and accountability with great trepidation. For this side equity and opportunity must be the central policy objectives, with the use of performance-based standards and accountability shaped specifically to achieve those ends.

These more fundamental differences help explain the seeming intractability of the controversy over standards-based reforms. They also serve as a caution to those who hope empirical studies of the impact of performance-based standards and accountability on disadvantaged students and schools will resolve it.

Nevertheless, in what follows, I attempt to bring to this argument the limited evidence we have about improving schools for low-income and minority students. I explore, from the perspective of this work, whether current proposals for curriculum standards and performance assessment are likely to foster such improvement. I also briefly suggest some empirical investigations that might increase the likelihood of formulating and implementing performance-based standards in ways that can minimize their potentially negative effects on disadvantaged schools.

WHAT DO WE KNOW ABOUT IMPROVING DISADVANTAGED SCHOOLS?

Unfortunately, the past three decades of efforts to improve urban schools have taught us more about what doesn't work to increase and sustain achievement in a challenging, "thinking" curriculum in disadvantaged schools than what does.

What seems not to work? First, big city school systems' efforts in the 1970s and 1980s taught us that highly prescriptive, specific, narrowly focused, rigidly sequenced programs (materials and/or instructional approaches) dictated by authorities removed from the school fail to appreciably improve either the conditions of disadvantaged schools or the achievement of low-income minority children, particularly with regard to complex curricular goals. (New York's Promotional Gates, Chicago's Mastery Learning, and Philadelphia's Curriculum Guidelines—all now defunct—provide good examples of such efforts.)

Precisely specified curriculum objectives, prescribed instructional scripts, and test-driven promotion criteria mandated and regulated by central offices may have contributed to some gains in younger disadvantaged students' basic skills (and decreased the black/white achievement gap in these areas). However, actual achievement gains were, for the most part, quite modest, and scores in most schools remained abysmally low.

Moreover, because this trend toward "improvement" reversed in the late 1980s, one suspects that earlier gains were as much a function of somewhat better social and economic conditions for low-income minority children and higher levels of federal and state education funding (including Emergency School Assistance Act funds to support desegregating school systems) during that period.

We also learned during this period that reforms based on a "let a thousand flowers bloom" approach (e.g., projects funded by Titles III and IV and, later, Chapter 2 of the Elementary and Secondary Education Act) also hold little promise for improving schools serving disadvantaged students. The considerable local innovations of the 1970s seemed to result most often in rich, rigorous, and creative programs in schools serving white and wealthy children and to institutionalize low expectations and misguided liberal dispositions to treat low-income minority children as "poor things" of whom too much shouldn't be demanded.

Perhaps because these reform efforts lacked a concerted, external pressure to establish new norms about disadvantaged children and what they can learn, these reforms too often exacerbated rather than ameliorated inequalities in opportunity and achievement.

What does seem to work? The "effective schools" research of the early 1980s (research consisting primarily of case studies of urban elementary schools) attempted to single out factors that explained why students at a small subset of disadvantaged schools scored higher on tests of basic skills than others. While that work has been criticized on both methodological (e.g., few controls) and substantive (e.g., actual gains were often quite small and not sustained over time) grounds, this work identified a set of school-climate factors that many effective schools shared: strong administrative leadership, an orderly environment, emphasis on basic skills, higher teacher expectations, and assessments tightly linked to instructional objectives.

However, accounts of these schools and subsequent efforts to replicate their successes were characterized most by the efforts of "heroic" principals who "turned around" deteriorating schools by subverting official policy and defying district rules rather than adhering to them. For all these reasons, as well as the fact that the narrow, basic-skills curriculum and tests of the effective schools are a far cry from the current emphasis on a "thinking" curriculum, this work has little to contribute directly to discussions of how large-scale policy development and implementation can promote high levels of achievement in complex curricula at disadvantaged schools.

More-recent evidence about effective approaches to improving disadvantaged schools comes from reports about a number of projects that have taken root in urban schools during the past decade. Again, much of this evidence comes from small-scale, action-research studies of individual projects, although some of these efforts have also been investigated as background for the congressional reauthorization of Chapter 1.

Of particular interest here are projects such as the following:

- Schools participating in James Comer's School Development Program, bringing together school staffs and social service providers in New Haven and elsewhere
- UCLA's collaboration with Lennox Schools (a port-of-entry community for Latino immigrants), whose strong, innovative parent involvement component has significantly enhanced literacy and language learning for language-minority students
- Baltimore and other city schools' intensive work with Johns Hopkins's prevention-oriented Success for All program
- Henry Levin's Accelerated Schools
- Learner-Centered Schools in New York City's District 4 and elsewhere, which work with Teachers College, Columbia University's National Center for Restructuring Schools and Teaching
- San Diego's countywide AVID program that has successfully integrated "unqualified," disadvantaged, minority high school students into college preparatory programs (and later college success)

- Philadelphia Collaborative's "charter" schools-within-schools that have brought about higher attendance and better grades at inner-city high schools
- CityWorks, a project-based, integrated academic and vocational high school program in Cambridge, Mass.

While the particulars of these projects vary in fascinating ways, what is most compelling for policymakers and practitioners is that they share a "deep structure"—a set of characteristics and principles that lie at the core of their work. In each of these projects we find the following:

- Inspiration—a powerful idea
- Charismatic leadership
- School-based organizational change and staff development
- Collaboration among teachers in new roles and relationships
- Supportive structures (e.g., small, heterogeneous communities of learners; interdisciplinary teams)
- Meaningful academic focus—rich, complex curriculum
- New conceptions of intellectual capacity, learning, and what *all* children can be expected to learn
- Strategies to increase students' confidence, effort, and persistence
- Emphasis on prevention rather than remediation
- Supportive relationships with parents and community

Importantly, these projects, taken together, represent an approach to improving disadvantaged schools that skirts the major pitfalls of earlier large-scale reform efforts. Schools' enthusiastic participation in these efforts follows from local flexibility with regard to curriculum and instruction rather than from policies that mandate prescriptive, narrowly defined, tightly controlled, externally developed programs.

However, none of these projects leaves the goals or direction of reform entirely to the discretion of local schools. Rather than "letting a thousand flowers bloom," local enactments are expected to be faithful adaptations of externally developed conceptual frameworks or models that are, in most cases, fairly well supported by research and theory. In all cases, participating schools must adopt powerful norms that all children can learn complex academic curricula when schools provide the supports and instructional conditions that make those curricula accessible to them.

This balance of external and internal influence is realized in most cases through schools' ongoing collaboration and identification with a project entity independent of any one school.

Research and practitioner experience with these projects and school improvement efforts generally suggest that such reforms are most likely to be developed and implemented successfully when policies provide the following "enabling conditions":

- Strongly articulated commitment to all children's achievement
- Professional "wiggle room"—permission to experiment and take risks
- Sufficient resources
- Experienced, knowledgeable, and skilled professionals
- A support scaffolding (e.g., university partnership; network) that stimulates change, builds capacity, and provides political clout

WILL PERFORMANCE-BASED STANDARDS LEAD TO EFFECTIVE POLICIES?

Nothing about performance-based standards per se prohibits policymakers or schools from developing these promising approaches to organization, teaching, and learning in disadvantaged schools. In fact, most advocates would see the norms and the techniques being employed by these projects as quite appropriate responses to these reforms and likely to compensate for special difficulties disadvantaged schools face in helping students achieve new standards. Moreover, advocates of performance-based standards reforms would probably argue that such standards coupled with school accountability will provide the motivation schools greatly need to adopt such strategies.

However, at least two widespread factors, not accounted for in performance-based standards reforms, make this result unlikely: severely limited capacity in disadvantaged schools, and strongly countervailing norms about the meaning and use of low-income minority children's performance on standardized tests. Evidence about the current distribution of schooling resources and practices and commonly accepted norms about the connections among race, social class, and intellectual capacity tell a disturbing story of systematic differences in students' opportunities to learn. It's most doubtful that performance-based standards reform will be powerful enough to counter these constraints.

Disadvantaged Schools Lack Capacity

Most schools with large low-income minority student populations suffer from serious fiscal, material, and human resource shortages—both absolute and relative to schools serving more privileged children. These shortages severely restrict disadvantaged schools' ability to make the changes that a successful response to performance-based standards reform requires—curricula based on big ideas in the content areas, solving complex problems, teaching for understanding and application of knowledge in new circumstances, hands-on, experience-based learning opportunities in the classroom, in the laboratory, and in the community.

Inequalities in resources and learning opportunities plague the U.S. educational system, with schools serving low-income minority students consistently having fewer. Jonathan Kozol addresses this issue quite dramatically in *Savage Inequalities* (1991). Some dismiss his book with charges of anecdotal evidence and inflammatory investigative reporting, but his assertions have been carefully documented in a number of other studies.

Most low-income and minority children attend schools that spend less. In some states per-pupil expenditures differ between neighboring high- and low-wealth districts by as much as a factor of three or more. Such inequalities persist even in many states where reforms have attempted to equalize resources. As a consequence, some students have less access than their more advantaged counterparts to well-maintained facilities, smaller classes, and equipment and materials. Such resources inevitably affect schools' ability to help their students develop academic and workplace competencies.

In addition to fewer resources, schools with large concentrations of low-income youngsters, African Americans, and Latinos typically provide fewer rich and demanding academic programs. My own recent study of the distribution of K–12 science and mathematics opportunities for NSF, *Multiplying Inequalities* (1990), shows that students in high-minority, high-poverty schools have limited access to rigorous courses, particularly the critical "gatekeeping" courses such as algebra in junior high school and calculus in senior high school.

Moreover, teachers at these schools place less emphasis on complex curriculum goals such as developing inquiry and problem-solving skills, and they offer fewer opportunities for students to become actively engaged in learning. Such differences extend beyond academic programs. The most recent National Assessment of Vocational Education found that schools with large disadvantaged populations also tend to have the weakest vocational programs.

Making matters worse, disadvantaged students have less contact than their more privileged peers with well-qualified teachers. High-minority and high-poverty schools more than others suffer more teaching vacancies, and principals have a tougher time filling them with qualified teachers. Students at affluent, white, suburban schools are far more likely to have teachers who are certified, hold bachelor's or master's degrees in their teaching fields, or meet qualifications set by teachers' professional associations, particularly in fields experiencing shortages such as math and science.

Such inequalities arise from complex educational and political conditions. Schools serving large concentrations of children from low-income families, African Americans, and Latinos often lack the political clout to command resources equal to those of other schools. Teachers often view these schools as less desirable places to teach, partly because of the economic and social disadvantages that shape their students' lives. These schools often pay teachers less than surrounding suburban schools and offer poorer working conditions (e.g., supplies, materials, physical facilities). As a result, these schools have far greater difficulty attracting and retaining well-qualified teachers. Some reject the significance of these findings, claiming that we have no evidence linking resources to student achievement. This claim rings increasingly hollow with recent studies showing that resources do indeed make a difference—particularly a combination of teacher expertise (gained through experience and advanced training) and smaller class sizes that permit teachers to employ more personalized instructional approaches.

An adequate and fair distribution of resources, programs, and teachers won't, by itself, guarantee that disadvantaged students will learn well. However, without sufficient funding channeled into high-quality instruction, policymakers and the public cannot expect much improvement from disadvantaged schools—regardless of how high we set new standards.

Not only do they have fewer resources, disadvantaged schools have less history and experience with such teaching and learning. This, too, diminishes their capacity. As noted earlier, schools serving disadvantaged students have concentrated during the past twenty-five years on "the basics," and, at least until the precipitous decline in support in the mid-1980s took its toll, these schools had some success in raising the scores of disadvantaged children on traditional basic-skills tests.

As a consequence of test score increases (however modest), the effectiveness and status of schools serving disadvantaged students have been gauged in just this way, and even with severe declines in schooling resources and social and economic conditions for poor children working against them, most schools serving disadvantaged students have directed their limited capacity toward meeting this standard.

Nothing about standards or performance-based assessment and accountability directly addresses these inadequacies, let alone provides for the resources and technical capacity for even highly motivated educators to create conditions necessary for improvement. We have no past experience that suggests that without serious, direct policy attention, these basic inequalities in capacity will be righted.

More and more, the public responds to low school performance with a withdrawal of support, frequently invoking the canard "don't throw money at the problem." Proof of improvement is often posed as the requisite condition for continuing, to say nothing of additional, support. But in good Catch-22 fashion, high performance is often taken as evidence that schools really don't need more after all. It is likely to take far more than new standards and assessments to bring children in disadvantaged schools equal access to basic conditions necessary to achieve.

Prevailing School Norms Close Down, Rather Than Open Up, Opportunities for Disadvantaged Students

In addition to limited capacity and experience, schools serving low-income minority students reflect prevailing norms about disadvantaged students' abilities that work directly against their provision of rich and challenging curricula and learning opportunities.

Partly because low-income and non-Asian minority students have historically scored lower than whites on large-scale achievement tests, schools typically judge them to be less able to learn complex knowledge and skills. To most policymakers and educators, then, it has seemed sensible, even just, that high-poverty, high-minority elementary schools accommodate their student populations with remedial academic programs that stress low-level topics and skills and that their high school counterparts offer programs skewed toward general academic and vocational curricula instead of more challenging, college preparatory work.

Even in racially mixed schools, this thinking has justified the overrepresentation of minority students in low-track classes and rationalized their absence from advanced courses and programs for gifted and talented students.

However, once labeled and placed in low tracks, students' *opportunities* to learn diminish significantly. Teachers typically provide remedial and low-level classes with little exposure to rigorous academic content, and they seldom ask them to grapple with material requiring critical thinking or problem solving. Low-track teachers less frequently design engaging, hands-on lessons, and they more often ask students to work alone—reading textbooks or filling out worksheets.

Such differences can't be justified as an appropriate tailoring of lessons to the needs of students at different ability levels. On the contrary, students in typical low-track classes simply have less exposure to instruction likely to help them become literate, critical-thinking citizens and productive members of a technological workforce.

These disadvantages may result, in part, from schools' tendency to place their least qualified teachers with low-ability classes. However, a powerful normative climate of low expectations for low-income minority children also limits the richness and rigor of the curriculum these schools and classrooms offer.

Within schools, many educators believe they base decisions about the distribution of curriculum and teaching on educationally sound criteria. For example, given an uneven teaching staff, high schools often decide that able students who are studying traditional college preparatory content need teachers with stronger preparation than less able students who are struggling to understand fundamental processes.

Of course, this assumption can be countered with the fact that unprepared teachers are least well equipped to diagnose students' learning problems and design activities to overcome them. However, most schools opt for a "rich get richer" pattern of allocating curriculum and teaching resources and opportunities. Politics play a part here, too. Parents of the highest-achieving children in the school often exert tremendous pressure to maintain curricular and instructional advantages for their children.

In U.S. schools, then, the normative responses to testing are clear: Large-scale assessments have consistently led to the closing down, rather than the opening up, of opportunities for disadvantaged and minority students. Judgments of "less able" and placements in low-level courses compound the problems posed by the uneven distribution of resources, curriculum, instructional strategies, and teachers that we find among schools. Low-income minority students are doubly disadvantaged: They attend schools that provide less and classrooms that expect less.

Schools' ability to mount programs that enable disadvantaged students to achieve the new standards requires new conceptions of intellectual capacity and learning, new understandings about how capacity and learning are manifest in different cultures, and norms that press schools to make knowledge accessible to all children as well as provide them access to a "standard" knowledge. While standards are likely to prompt new norms, these are norms that define content and high levels of performance vis-à-vis that content (i.e., curricular norms).

For example, standards may assert that all children should demonstrate high levels of performance in a complex curriculum in order to succeed personally in the workforce of the future and to contribute to the quality of that workforce. They will not, however, alter fundamental norms that currently support narrow conceptions of learning ability, its bell-shaped distribution, and its uneven distribution by race and social class.

Without such norm shifts, new standards are unlikely to persuade the hearts and minds of well-intentioned educators that, even though low-income minority students *should* reach complex curriculum standards, they either *can* or *will*.

In spite of schools' good intentions, then, their differentiation of curriculum and instruction—supported by past and current testing—most often widens the achievement gap between students judged to be more and less able.

Without a focused effort to alter the norms that underlie these practices, a dramatic increase in the emphasis on assessing students' performance vis-à-vis new standards will undoubtedly exacerbate such inequities.

Without Fundamental Normative Shifts, Can Performance-Based Standards Reform Foster Professional Will?

Finally, even many advocates of systemic reform argue that it's unlikely that few, if any, of the prospective sanctions that could be levied against low-performing schools would be powerful enough, in themselves, to motivate professionals in disadvantaged schools to ratchet up the quality of their curriculum and instruction sufficiently.

Jennifer O'Day and Marshall Smith (1993) rightly wonder in their paper on equity and systemic reform, "What the devil do you do?" Most really potent actions simply are not feasible. Are we likely to take a school's or district's money away? Replace all of the old teachers and administrators with new ones? (As if there are the human resources available.) Close the place down? Taking such extraordinary actions toward one school is complex and difficult, as some recent state takeovers of bankrupt schools have demonstrated. Establishing such measures as the large-scale policy response to the typical performance of low-income minority schools is mind-boggling. Further, despite considerable evidence that educators do pay more attention to those things that are the subject of evaluation, the rather blatant carrot-and-stick approach being contemplated in many quarters relies on motivational models derived from individual-based, behavioral psychology. Such models are not only increasingly discredited as valid and useful explanations for individual learning and actions (notably, by many of those involved in developing standards), but as Seymour Sarason (1982) has often noted, they have little hope of being transferable to the level of complex institutions.

ARE THERE ALTERNATIVES?

What correctives might improve the prospect of standards reforms in disadvantaged schools? The most popular and salient alternative scenario or corrective is the inclusion of school delivery or opportunity to learn (OTL) standards in any system of standards and accountability. The National Council on Educational Standards and Testing (1992) defined such standards as "criteria to enable local and state educators and policymakers, parents, and the public to assess the quality of a school's capacity and performance in educating their students in the challenging subject matter set out by the content standards."

Interestingly, opportunity to learn has been part of the thinking in mathematics education reform for some time. It first emerged when Edward Kifer analyzed the achievement of students who participated in the Second International Mathematics Study by relating students' answers to particular items to whether or not students in various countries had actually been exposed to the knowledge required to answer the test items correctly.

In an extraordinary report, *The Underachieving Curriculum* by Curtis McKnight et al. (1987), Kifer reports the astounding if commonsensical finding that students in various countries learn the mathematics topics and skills that they are taught in school, and they don't seem to learn those topics and skills that are not a part of their mathematics curriculum. This study helped refocus reformers' thinking about the importance of holding schools accountable for what they teach as well as holding students accountable for what they learn.

Why might OTL standards help disadvantaged schools? O'Day and Smith (1993) argue that the supportive infrastructure and resources that systemic reform can foster provide the best hope for extending reforms based on challenging curriculum and complex instruction to disadvantaged schools. However, even they do not expect that, in the absence of explicit policies, systemic reforms composed of content and performance standards alone will be implemented "fairly" across schools.

O'Day and Smith argue that other standards are necessary to provide the "operational specifications" for assessing whether schools are providing students with the opportunity to learn the content and skills in the content standards. OTL standards would function to (1) amplify the vision and structure provided by content standards, and (2) establish the conditions under which students could be legitimately held accountable for their performance on assessments of content standards.

Others, including myself (in a 1989 piece on educational indicators), argue that policymakers, educators, and the public can only gain much-needed insight into why students at a particular school score as they do when reports about student performance are "contextualized" with information that OTL standards would provide—the resources, processes, and activities in which scores were achieved. Moreover, this information may be essential to counter schools' temptation to artificially elevate their scores by "pushing out" students who don't score well—through special education identification, grade retention, or encouraging them to leave. Similarly, knowing that students' performance will be interpreted in the light of OTL may help discourage teachers from gravitating toward schools where students are easier to teach. Finally, these standards can serve to monitor the state's basic responsibility to treat well students who are compelled by law to attend.

Two broad categories of opportunity-to-learn standards are proposed most frequently: resource standards and practice standards. Resource standards would include fiscal, human, and material resources and address such questions as the following:

- Do schools have equitable access to the *dollars* necessary to buy what's needed to provide students with access to the content standards?
- Do schools have *fully qualified teachers* with content knowledge and knowledge of how diverse students with diverse backgrounds, abilities, and learning styles develop and learn?
- Are there *curriculum materials, equipment, and laboratories* available for learning the content standards?

Proposed practice standards range from those quite directly related to teaching and learning of content standards (such as teachers' knowledge of the curriculum they're teaching) to broader conditions essential for high-quality teaching and learning more generally. While no two analysts agree completely, practices standards are expected to address the following types of questions:

- Is the implemented curriculum rich and challenging, and does it match the standards?
- Are pedagogical strategies powerful and developmentally appropriate, and do they build on the various strengths of the students?
- Do professional development activities build capacity to enact the standards?
- Is the school environment safe and respectful for adults and students?
- Are there learning environments that encourage close, sustained relationships between teachers and learners?
- Do school environments convey confidence in children's intellectual capacity?
- Are there requisite conditions for professional development and growth?
- Are parents and the community meaningfully involved?

Proposals for OTL standards have generated considerable controversy. For example, it was only after much heated debate that the House Education and Labor Committee attached amendments that provide federal support and oversight for states' development of such standards to the Clinton administration's Goals 2000 reform bill. Not long after, the *Los Angeles Times* (1993) noted that state legislatures were responding with consternation to this news: "Local officials fear this requirement could become a back-door federal mandate to equalize all education spending, or could spur lawsuits from parents whose children attend schools that don't meet the federal standards."

Indeed, developing and implementing a federally supported set of OTL standards would require contentious political debates about what exactly constitutes "opportunity" and what criteria should be used for judging whether or not students in various schools have access to a fair share (which is not necessarily an equal share) of relevant resources and experiences.

As suggested above, assessing simply whether particular resources and opportunities exist doesn't go far enough. We also need to know, Are there enough? Are they of sufficiently high quality? Are they appropriately used in the context of learning? Thus, fears about a redistributive impact of such standards are not unfounded. That's the point!

Others have worried that OTL standards will be so narrowly construed and rigidly prescriptive that they will repeat the mistakes of highly regulated programs in the past. Detailed, cumbersome mandates for particular resource configurations and particular practices would cripple local flexibility and divert attention away from what students actually accomplish.

Others argue that the substantive and technical problems of measuring OTL are insurmountable. Analysts working on developing valid and feasible measures of those aspects of opportunity most proximal to students' engagement with curriculum and instruction—such as the enacted curriculum—are only cautiously optimistic about their efforts.

Finally, even supporters of OTL standards worry that school-level measures will mask variations in the distribution of resources and opportunities *within* schools that parallel between-school inequities—low-income and minority students having less access as a consequence of test scores and other "evidence" that they are less capable.

HOW COULD OTL STANDARDS TRIGGER IMPROVEMENT?

As important as these political, substantive, and technical concerns are, the essential question here regards the impact such standards are likely to have on the capacity and will in disadvantaged schools. Will OTL standards make any difference? If so, under what conditions?

At the policy level, regularly assessing and reporting authoritative, widely agreed-upon resources standards will make a difference to the degree that they make glaring inequities among schools salient in public discussions about how well students at various schools are doing, and whether or not it's appropriate and fair to make high-stakes decisions about students based on their test performance.

Public disclosure lies at the heart of accountability in a democratic system, whether the news is good or bad. But public disclosure requires more than just publishing data; it also requires making reports understandable to the various publics concerned about schools—policymakers, educators, parents, the media, and the general public.

Effective public disclosure requires devoting substantial time and energy to placing student performance data in the context of opportunities to learn and to helping various audiences to interpret these contextualized outcomes. In contrast to an exclusive focus on student performance, specific information about the dollars going to particular schools and how those dollars are spent can inform and shape state and district policies that support improvement at disadvantaged schools—for example, resource generation, resource redistribution (such as teacher assignment policies), capacity building, and staff development.

At the level of practice, matters become far more complicated. Promoting a standards-based improvement process in local schools necessitates new, more radical strategies than what states and districts typically know how to do or see as their traditional role. How the information is gathered about the opportunities a school provides and how the information is fed back into the school is likely to make a huge difference in whether and how it is used to spur increased capacity and opportunities.

Simply collecting and reporting to schools basic lists of the amounts and types of resources they provide and practices they engage in (and comparing these to "standards") will not be effective. Even highly psychometrically sophisticated measurement and reporting strategies will fail to have their desired impact if the way they're used alienate those in schools who are best suited to act on the results.

If information about where schools stand in relationship to new standards is to spur school improvement, we need mechanisms for reporting assessment information back to schools in ways that educators can use to inform their practice. Carrot-and-stick approaches might provide some motivation, but they cannot, in and of themselves, increase capacity for change. Ideally, an implementation strategy will pay conscious attention to normative and political, as well as technical, issues related to educational change.

What might such a process be like? Several analysts have proposed school-based methods for regularly eliciting and analyzing information about practices. For example, Ken Sirotnik and I offered in 1986 a "critical inquiry" process led by a critical friend. Linda Darling-Hammond proposed in 1992 that peer study-teams engaged in elaborated types of accreditation reviews. And, in 1993, William Clune suggested regular site visits and review of school portfolios by a professional inspectorate.

These proposals share an assumption that assessments of OTL are unlikely to produce unequivocal information; most of the data will be susceptible to various interpretations. Consequently, whatever the specific mechanisms employed for the collection and reporting of school-level data, their impact is likely to depend on the extent to which they include an ongoing process of reflection on those data—methods for deliberating about the data's meanings and their implications for changing current practices.

These proposed strategies also suggest that schools are unlikely to be able to make sense of these data and apply them to their experiences without support, encouragement, and even friendly pressure from the outside. At its best, such inquiry among educators at a school could actually mirror effective approaches to teaching a complex "thinking" curriculum, wherein small heterogeneous communities of learners solve problems that challenge conventional ways of thinking, acting, and relating to one another in school.

In this respect, the most promising strategies for standards-driven school improvement may be indistinguishable from good schooling itself.

BENEFITING ONLY THE DISADVANTAGED?

Will such efforts benefit *only* disadvantaged students and schools? Probably not. While the inequities in U.S. education are large and destructive, perhaps an even larger problem is the general lack of public will to support education. OTL indicators would bring into public view and inform policy debates about the impact of the widespread decline in education resources. No one doubts that teachers and students in white and wealthy suburban schools are far better supported than their counterparts in disadvantaged schools.

Yet, this year, for example, affluent white teenagers in at least one property-rich school system outside of Los Angeles are crowded in groups of about forty into their high school classrooms. Facing five such classes each day, teachers wonder how they can possibly offer authentic, personalized instruction; just learning their names is a challenge.

Fortunately, the classrooms are in good repair, and there are plenty of textbooks to go around, even if most were written long before California's new curriculum frameworks refocused the subject areas. Unlike in many nearby systems, all of their teachers are well qualified, even though professional development opportunities are increasingly hard to come by. And, while rather dramatic pay cuts have seriously undermined faculty morale in neighboring Los Angeles Unified School District, these teachers have found an absence of any pay increases during the past several years somewhat easier to swallow. Ballot measures proposing additional resources for this school system, as highly regarded as any, have been turned back year after year.

Conditions in these schools constrain the learning opportunities of these most advantaged students. In the absence of OTL standards, new content and performance standards may mask these grim realities and shift the burden of responsibility even farther away from the public. Even in affluent communities, deliberations about what conditions are necessary to support students' right to achieve and where the responsibility lies for providing those conditions can be aided by carefully conceived and reported information about students' opportunities to learn.

WHAT RESEARCH MIGHT THROW LIGHT ON THESE MATTERS?

Clearly, all of the above raises more questions than it answers—questions that are appropriate for research. However, one productive strategy for action research is inquiry in sites where current efforts incorporate elements of standards-based reforms. For example, California's curriculum frameworks and the California Learning Assessment System (CLAS) assessment, in effect, set guidelines for teaching and learning in a number of subject areas that are quite similar to those being proposed in the national standards-setting projects. Adoption of the frameworks is not mandatory for California schools, but the informal pressure to develop local frameworks-based curricula is great. As yet, students' scores on California's state assessment do not inform high-stakes decisions, but school scores are published widely and community pressure to do well is great.

New York provides a longer-standing example of state mechanisms for establishing and linking curriculum goals and assessment. Even though the Regents course syllabi and examinations are not required for all students, they drive much of the academic curriculum in middle and senior high schools. Moreover, high-stakes decisions about students (type of diploma earned, college entrance) follow from their scores.

These states provide useful policy contexts for investigating the impact of standards-like policies on disadvantaged schools:

- Do these mechanisms seem to spur institutional efficacy and individual motivation in these schools? Trigger efforts to increase capacity? Alter norms about what disadvantaged students can and should learn?
- How does an emphasis on content and student performance relate to considerations of resource distribution and students' access to knowledge and various instructional opportunities?

Similarly, efforts to understand the process of developing and implementing OTL standards and their impact on schools might be undertaken in settings where similar processes are being tried. One good place to undertake this work at the senior high school level is Massachusetts, where the recently passed Massachusetts Education Reform Act mandates that the Department of Education publish annually school and district profiles.

In anticipation of this new policy, the department has developed a report template and initiated training to assist high schools in compiling locally generated assessment reports as an alternative to state-generated profiles. The template is particularly relevant to OTL standards, since it incorporates "enabling conditions"—students' access to knowledge, the environmental press for achievement, and professional working conditions—as well as background and outcome data. Its format allows for narrative information to supplement and support statistical data.

Schools participating in this state-sponsored, school-based effort should provide fertile ground for action research on the impact of performance-based standards reforms that include OTL standards on disadvantaged schools.

New York state provides a second interesting opportunity. The National Center for Restructuring Schools and Teaching, which is directed at Teachers College, Columbia, by Linda Darling-Hammond and Ann Lieberman, is experimenting with training teams of educators to serve as reviewing teams for a school quality review process that would fit with that state's new Compact for Learning. This effort engages schools not only in collecting information about curriculum, opportunities, and outcomes but also in consultation, joint planning, collegial observation, and shared inquiry.

Consequently, this project could be quite well positioned to explore the processes whereby information about disadvantaged schools vis-à-vis standards can be collected and reported in ways that increase school capacity and commitment to improvement. Similar efforts might be undertaken in conjunction with California's newly revised requirement that schools undergo a periodic Program Quality Review.

CONCLUSION

In many states, the vision of standards embedded in systemic reform has already emboldened policymakers to take as a serious policy objective the idea that all children can learn challenging and complex curricula and demonstrate that learning on wide-scale assessments. Witness the number of states and large school systems that have signed on to the New Standards Project (nineteen and four, respectively, at last count).

These reforms, however, undoubtedly require more than widespread agreement with the vision, particularly if they are to make fundamental changes in schools that serve disadvantaged children. They will likely need an equally visionary and bold implementation strategy that departs substantially from traditional mechanisms such as mandates, narrowly construed incentives, or standardized technical assistance. Creative approaches to collecting and reporting information about whether schools provide students with the resources and opportunities they need to achieve new standards are likely to be a part of such a strategy.

However, OTL standards' contributions to improved schooling for disadvantaged students will depend on the thoughtfulness with which they are designed, reported back to schools, and applied to problems. Neither simple checklists nor complicated regulations will do. In the end, standards can be no more than tools to aid an ongoing political dialogue about what we want from our schools and how those ends can best be accomplished. They will not carry with them a single interpretation of past practices, offer clear judgments about present conditions, or provide direct answers about what should be done next.

All the same, they may bring new knowledge to bear on educational issues, stimulate more thorough discussion and debate, and suggest creative new solutions to problems. If OTL practice standards are allowed to serve these more modest purposes, they may indeed prompt far better opportunities and achievement for students in currently disadvantaged schools. Thus, it behooves those interested in standards reforms to pursue the conditions under which they will be salient and legitimate to educators.

· · · · · · · · · · · · · · ·

NOTE

This article was originally prepared as a background paper for a Research Forum of New Standards Project on "Effects of New Standards and Assessments on High Risk Students and Disadvantaged Schools" held at Harvard University, October 1993.

· · · · · · · · · · · · · ·

REFERENCES

Clune, William H. "The Best Path to Systemic Educational Policy: Standardized/Centralized or Differentiated/Decentralized." *Educational Evaluation and Policy Analysis* 15 (1993): 233–54.

Darling-Hammond, Linda. "The Implications of Testing Policy for Quality and Equality." *Phi Delta Kappan* 73 (1991): 220–5.

____. "Inequality and Access to Knowledge." In *The Handbook of Multicultural Education*, edited by James A. Banks. New York: Macmillan, 1995.

____. *Standards of Practice for Learner-Centered Schools*. New York: National Center for Restructuring Schools and Teaching, Teachers College, Columbia University, 1992.

Kozol, Jonathan. *Savage Inequalities*. New York: Crown, 1991.

Los Angeles Times. "State Lawmakers Feel New, Closer Bond to White House but Same Old Anxiety." 31 July 1993, p. A15.

McDonnell, Lorraine M. "Assessment and Equity in Educational Reform." Paper presented at the annual conference of the National Center for Research on Evaluation, Standards, and Student Testing. Los Angeles, 1993.

McDonnell, Lorraine M., Leigh Burstein, and Tor Ormseth. *How Can We Tell What Schools Really Teach?* Santa Monica, Calif.: RAND, 1991.

McKnight, Curtis, F. Joe Crosswhite, John A. Dossey, Edward Kifer, Joan O. Swafford, Kenneth J. Travers, and Thomas J. Cooney. *The Underachieving Curriculum: Assessing U.S. School Mathematics from an International Perspective.* Champaign, Ill.: Stipes Publishing Co., 1987.

National Council on Educational Standards and Testing. *Raising Standards for American Education.* Washington, D.C.: The Council, 1992.

New Standards Project. *The New Standards Project, 1992–1995: A Proposal.* Pittsburgh: University of Pittsburgh, Learning Research and Development Center, 1992.

Oakes, Jeannie. *Improving Inner-City Schools: Current Directions in Urban District Reform.* Santa Monica, Calif.: RAND, 1987.

_____. *Keeping Track: How Schools Structure Inequality.* New Haven, Conn.: Yale University Press, 1985.

_____. *Multiplying Inequalities: The Effects of Race, Social Class, and Tracking on Opportunities to Learn Mathematics and Science.* Santa Monica, Calif.: RAND, 1990.

_____. "What Education Indicators? The Case for Assessing the School Context." *Educational Evaluation and Policy Analysis* 11 (1989): 181–99.

Oakes, Jeannie, Molly Selvin, Lynn Karoly, and Gretchen Guiton. *Educational Matchmaking: Academic and Vocational Tracking in Comprehensive High Schools.* Santa Monica, Calif.: RAND, 1992.

O'Day, Jennifer, and Marshall Smith. "Systemic School Reform and Educational Opportunity." In *Designing Coherent Educational Policy,* edited by Susan Fuhrman. New York: Jossey-Bass, 1993.

Porter, Andrew. "School Delivery Standards." *Educational Researcher* 22 (1993): 24–30.

Resnick, Lauren. "Assessment and Equity in Educational Reform." Paper presented at the annual conference of the National Center for Research on Evaluation, Standards, and Student Testing, Los Angeles, 1993.

Sarason, Seymour. *The Culture of the School and the Problem of Change.* Boston: Allyn & Bacon, 1982.

Sirotnik, Kenneth A., and Jeannie Oakes. "Critical Inquiry for School Renewal." In *Critical Approaches to the Organization and Improvement of Schooling,* edited by Kenneth A. Sirotnik and Jeannie Oakes. Boston: Klewer, 1986.

Mathematics Excellence for Cultural "Minority" Students: What Is the Problem?

Asa G. Hilliard III

THE school achievement record of certain cultural minority groups, specifically African Americans, American Indians, and some Hispanic groups, is widely known to be a poor one. Overwhelmingly, mathematics educators appear to be perplexed; few seem confident in their ability to teach these students. There are many other cultural or ethnic groups in the schools, some experiencing the problem of low performance, others appearing to have no problem and even appearing to excel. For those who excel in spite of cultural or ethnic differences from the mainstream, it is obvious that culture is not a problem.

Habitually, educators identify students in a variety of ways; "race," class, and other ethnic or cultural identities are ascribed by educators or asserted by students and their parents. "Race" is mainly a political category used to refer to phenotypical variety. There is no empirical evidence that race has any real meaning for teaching and learning other than its political meaning.

Politically, the question is one of equality of treatment by school personnel for all students. Contrary to popular belief, the pedagogical issues associated with race have nothing whatsoever to do with student learning capacity.

Socioeconomic status or social class is also a political-economic category and, like race, has no significance for teaching and learning other than its

Asa Hilliard III is Fuller E. Callaway Professor of Urban Education at Georgia State University. He holds joint appointments in the Department of Educational Policy Studies and the Department of Educational Psychology/Special Education. He is a teacher, psychologist, and historian.

political-economic meaning. The learning capacities of poor children are intact, and if impaired, are not so because of their social class so much as because of the treatment they receive on account of their social class.

Cultural or ethnic identity is meaningful for teaching and learning simply because culture is a powerful aspect of the *context* for teaching and learning. All students are nurtured in a cultural environment and "belong" to it. Some of these cultural or ethnic groups are highly distinct or unique. Others have considerable overlap with other cultural or ethnic groups or with "the mainstream."

CULTURE IS MEANINGFUL

Students differ in terms of the strength of their attachment to particular cultures, including their historical culture of origin. We must recognize that both cultural or ethnic groups and their respective individual members are constantly changing—sometimes not in the same direction. As a result, we should not be surprised to learn that individuals may be socialized in an entirely different cultural environment from their culture of origin. This is why phenotype is not a guarantee of cultural identity.

For the educator, however, culture is meaningful because it hints at what we can take for granted in terms of common information, language, priorities, and possibilities for rapport and communications both between teacher and students and among students. It tells about preferences and styles of information processing as well as ways of socialization (Hall 1977).

Friction and conflict among cultural and ethnic groups are not automatic just because of diversity, nor are problems in teaching and learning automatically due to diversity. Cultural and ethnic diversity may be manipulated and exploited by those with a political agenda (Weinberg 1977); however, this aside, diversity can and ought to enrich the school environment.

For most students (and teachers?), mathematics itself is a foreign language and a new cultural experience. In fact, it has been observed frequently that traditionally low-performing cultural minority students do better on standardized, quantitative reasoning tests than on that part of standardized tests that purports to measure verbal skills or general information.

As the performance of cultural or ethnic minority students shows, far from being inaccessible to minority students, mathematics, properly taught, can be a powerful tool for raising student self-esteem (Egan 1992; Douglas 1981; Cummings 1978; Barrett 1992). This is so because of the cultural value in the wider society that is placed on mathematics achievement in relationship to other subjects.

Students, too, feel that there is a greater chance that their true achievement will be recognized in mathematics than in those areas of the curriculum more susceptible to teacher judgment. The standard cultural or ethnic categories of African, European, Asian, Hispanic, Indian, and so on are quite gross. Subgroups within each of these categories are meaningful ethnic or cultural groups. For the mathematics teacher, however, two topics are of primary importance for consideration when responding to ethnic or cultural diversity. On the one hand, the ethnic or cultural content of the mathematics curriculum is important. On the other hand, the process of teaching and learning is also important. Both are matters directly related to what teachers can accomplish in mathematics.

All cultural or ethnic groups are part of traditions that have created mathematics and have contributed to the creation of our common mathematical knowledge base. The ancient cultures of Mexico, Africa (including Egypt and the Ishango in Zaire), and Chaldea are cases in point (Diop 1991; Zaslavsky 1994). In addition, as ethnic or cultural populations have merged over time, individuals from a wide variety of cultural groups have contributed to the advancement of mathematics.

Generally, it is hard to convey this information using popular American textbooks in mathematics. Either curriculum content should be cleansed of all cultural and ethnic references, such as the names of individuals or references to the national origin of mathematicians and mathematics, or care must be taken to ensure that the legitimate role of diverse cultural and ethnic groups and individuals in the history of mathematics is highlighted. The failure to do this will convey a false impression to students that some cultural or ethnic groups are outside the mathematics tradition therefore and, by inference, are less capable as learners. Teachers have a responsibility to guarantee attention to the work of ethnomathematics.

SENSITIVITY TO CULTURAL STYLES

The teaching and learning process can be enhanced by those teachers who have appropriate sensitivity to cultural styles in information processing, not only to be compatible with family cultural patterns but as a source of alternative ways to present to any students an enriched mathematics learning environment (Hilliard 1992, 1989).

There is great and growing cultural or ethnic diversity among students, but as Bill Johntz, the great mathematics teacher and founder of the national Project SEED in Berkeley, Calif., has shown, mathematics can be a culture that provides the opportunity for diverse student populations to come together in harmony as they enjoy the beauties of nature through the language of mathematics ("Seed" 1968; "The Common Language" 1970; "Higher Math" 1971).

Professor Abdulalim Shabazz, chairman of the mathematics department at Clark Atlanta University, is almost single-handedly responsible for the education of more than half of the African Americans who hold a Ph.D. in mathematics. He is fond of saying:

> Mathematics is life and life is mathematical. Mathematics is inherent in our experience. Rhythms and patterns are there and are experienced by all, some of whom fail to see the mathematical significance of them. No people is alien to mathematics. However, some are unconscious of their relationship to it. It is the role of teachers to bring to the level of consciousness the beauty of mathematics as both a pragmatic and an aesthetic experience. (Hilliard 1991)

Explanations abound for low student achievement. Some educators cite cultural deficiency, language impairment, low intelligence, the culture of poverty, family background, peer pressure, and a host of other reasons. As these reasons are identified, educators often express the idea that the remedy is beyond the control of teachers, and that while some educators may agree that some of the students are mentally capable of higher levels of achievement, even these few students will probably fail, due to the overwhelming magnitude of the environmental problems that they face.

Many activities are under way to improve academic achievement for low-performing students. Hundreds of educational researchers are busily investigating the problem. In addition, professional development schools (a fad) are placed within the neighborhoods of low-performing students. There, university personnel and public school teachers, as well as community members, are joined together to experiment and to find a solution to the perennial "problem of low cultural minority group achievement."

Rarely do we hear of success in producing academic achievement for African, Native American, and Hispanic students. When we do hear of such achievement, it is trumpeted as a miracle, as the exception that proves the rule, as the work that can only be done by teachers who have a special charisma, as outliers that have to be regarded as statistical errors or mere accidents.

Ron Edmonds (1979) raised a powerful question:

> How many schools would you have to see that are getting excellent achievement results with traditionally low-performing students before you would be convinced of the educability of the children and the power of schools? If your answer is more than one, then I suggest that you have reasons of your own for believing that the success of children derives more from family background than from the schools that they attend.

Edmonds was right. Major teacher education institutions and large staff development programs do not commonly boast of their ability to produce excellence-level academic achievement with African, Native American, and Hispanic students, especially when the students are also poor. Moreover, few of those who are responsible for the education of teachers appear to be aware of the existence of teachers in schools who routinely produce academic excellence with these very students (Backler and Eakin 1993).

One would expect that national, state, and local agencies that are responsible for the education of America's children—and especially those who are responsible for staff development—would have a repertoire of success experiences with these students. Otherwise, what does it mean to be a specialist, if one is ignorant of best practice?

The author began his career in the public schools in Denver, as a mathematics teacher at the secondary level. His first assignment was in a low-income area serving primarily poor white and Hispanic students. He offered his seventh-grade class an opportunity to learn advanced mathematics, promising them that if they would come to school for an hour before school began, and if they would maintain perfect attendance, that he would teach a course in algebra that normally was not offered until ninth grade.

This was in 1955 and 1957. Not only did the students accept the challenge, they showed themselves to be eager and apt, easily mastering the mathematics intended for higher-level students. The author was convinced then, as now, that most educators underestimate the intellectual capabilities of all students—particularly those who traditionally have a record of failure.

There were no secrets, no special techniques, there was no mystery. One thing was certain: The children could not learn algebra if they were not exposed to it.

SEEKING THE CAUSES OF LOW ACADEMIC ACHIEVEMENT

The evidence was clear then as now that neither the child nor his or her culture were impediments to learning. Yet the professional literature is packed with references to pinpoint the child's intellectual abilities (Snyderman and Rothman 1988), socioeconomic and family background, and culture as the primary causes for low academic achievement.

Consistent with these beliefs is the professional practice that has emerged to respond to the perceived problem. Psychological assessments, mainly IQ testing, are developed, refined, and applied, with no abatement of this practice.

Studies are made of the child's socioeconomic status, and social welfare services are provided. Cultural studies of all types are performed, focusing on language, values, style, and other aspects. In short, there continues to be a varied, massive data-gathering effort, focusing on the child, family, and community—largely nonschool cultural factors—to explain low achievement.

There have been more-recent studies of the school environment. The Effective Schools movement is the best example of a broadly based, sustained attempt to understand how schools can succeed. While some success is reported, the results may be regarded as minimal. Typically, schools are seen as successful with cultural minority groups when they achieve more than one grade level per year or when the students are brought up to grade level. Of course, this assumes that grade level is an appropriate standard for any child.

As we examine reports from around the nation, we do not get a sense of any real momentum for improvement of the education for cultural minority students, especially in mathematics. Isolated examples of success are reported, mainly in the mass media. However, the professional literature does not provide much information about successful mathematics instruction institutions, especially information about excellence in mathematics achievement for cultural minority children.

Because of our general experiences and beliefs, there are several issues:

1. Mental measurement for cultural minority groups
2. Learning styles for cultural minority groups
3. Language development and diversity for cultural minority groups
4. Special education and cultural minority groups

Since the author's early teaching years, he has sought out the teachers and schools that are successful in producing the highest levels of achievement in mathematics with African, Native American, and Hispanic students, mainly because of the almost universal belief that these students are difficult to teach.

The empirical reality of success under widely varying conditions all over the United States is a challenge to the pessimism of educators. It seems that some teachers attempted to cover their own lack of skill by, consciously or unconsciously, fabricating explanations for student failure. It seems also that some educators, especially in the area of mathematics, derive ego satisfaction from being identified with a "hard" subject.

The more mysterious and obscure mathematics can be made to appear, the greater the accomplishment of the mathematics teacher may seem to be, because he or she has mastered this esoteric material.

The time has come to face reality. *The simple reality is that some teachers succeed in producing high student achievement, no matter what the students' background, ethnic or economic.* Look at the successes that many teachers have in producing academic excellence with low-income African American, Native American, and Hispanic students (Backler and Eakin 1993; Hilliard 1991). We can look at the successes, examine the commonalities among successful schools, and ask, What prevents the unsuccessful teachers from becoming successful? How can we make all schools as successful, given ethnic and cultural diversity?

ASSESSMENT AND CULTURAL MINORITY STUDENTS

Given the fact there are many examples of high levels of student achievement that are not predicted by traditional assessment techniques, nor attributable to the use of such techniques, we must gain a clear perspective on the place of assessment in education.

Traditionally, the most popular forms of school assessment are standardized achievement tests and standardized intelligence or IQ tests. A separate set of issues is associated with each type of testing. Then, of course, the extent to which testing or assessment is itself an issue is important, for we may test without assessing; we may even assess without testing.

A classic statement on assessment for school purposes has been made by Grant P. Wiggins (1993). His state-of-the-art statement properly connects testing activities to the larger function of sizing up a total educational situation, in order to design effective instruction. Testing is a more limited, though important, aspect of that enterprise.

Achievement testing in mathematics is important, and its central issue is content validity, the match between the actual coursework taught and the content of the assessment instrument. Empirical examinations of the relationship between coursework and standardized tests have shown consistently that there tends to be a gross mismatch between the two. If there is an area of curriculum where an excellent match should be expected, it is mathematics.

For many educators, the prospect of reaching agreement on course content objectives is greater in mathematics than it would be, say, in social studies. However, as was shown in research at Michigan State University's Institute for Research on Teaching (Hilliard 1982), there is most often nearly a 50 percent mismatch between course content as indicated in textbooks and standardized tests of mathematics achievement in the elementary grades.

This measurement problem creates a real assessment problem: It is hard to know the meaning of either a low or high score given such a mismatch. Of course, the problem can be corrected by aligning the curriculum to the test content, by changing the test, or by creating content-valid tests based on local curricula. But the status quo cannot be tolerated.

Intelligence or IQ tests are also used in schools, mainly to provide the basis for guessing about future levels of performance for students—prediction. Here, content validity is not the issue. Predictive validity is. There is a small association between IQ scores and later academic achievement for students; however, for the mathematics educator to use these is to overlook what we have learned from observing good teachers.

Good teachers are able to get their students to achieve without reference to IQ tests and in spite of predictions made based on them. In fact, some of the most extreme critics of IQ tests have argued that they have no value in the education process at all (Hilliard 1994).

Relying on IQ-test results causes teachers to develop erroneous expectations about student capacity. We also know that low expectations by teachers are associated with low efforts by teachers to assist students to overcome difficulties in learning; therefore, many educators feel that the use of results from tests of intelligence by teachers may actually cause low performance.

While a valid achievement test in mathematics may be of value to teachers, the IQ testing of students has yet to demonstrate its value. But neither type of testing provides sufficient information for teachers to design better instructional practices. Available to teachers is another form of assessment that can be of great benefit to students: the assessment associated with applied cognitive psychology (Feuerstein 1980, 1979; Lidz 1991, 1987).

The purpose of the third form of assessment, dynamic assessment, is to look beyond learning products to learning processes to determine whether dysfunctional learning processes are impeding students in their acquisition of mathematics material. More than thirty years of work has gone into the development of one approach to applied cognitive assessment and mediation, sometimes referred to as remediation. That is the work by Reuven Feuerstein and his associates, but there are other approaches as well.

The real question is what role assessment does, can, or should play in improving mathematics achievement among students. The evidence seems to be that IQ actually serves no function as far as instructional improvement is concerned. Achievement testing may serve a limited function, but if assessment is to serve a powerful function, then it must include the assessment of the teaching and learning process, and link that to powerful mediational strategies as well.

Finally, we must recognize that many teachers are able to produce achievement without reference to formal assessment processes at all. This means that mathematics achievement can be realized by more than one pathway.

THE CULTURAL ISSUE

It is obvious to any casual observer of education in the United States that cultural diversity is the norm. The U.S. population is made up of people originating from all over the world, and, if anything, that diversity is increasing. While some educators see cultural diversity as the source of problems in educational achievement, cultural diversity itself may not be the culprit.

As far as education is concerned, the specific parts of culture that are of most interest have to do with language, communication styles, experiential repertoire (including general information and vocabulary), and related items. Students and teachers may be matched or mismatched on these aspects of culture. The question is, what effect the match or mismatch in culture among students—or between students and their teacher—has on the achievement of children.

Those who study culture and education professionally are able to demonstrate that the essence of culture is that it is learned and therefore that it can be taught (Hall 1977; Spindler 1987). No one can deny the range of cultural diversity that exists—but no one can deny the capacity of learners to absorb new cultural material. The question is whether educators will allow the time and exposure necessary to acquire a common cultural core, to assist in building an educational environment that is comfortable for all students.

Too frequently, cultural diversity is perceived as a pedagogical problem. In fact, it is of more consequence as a political rather than as a pedagogical issue. It is indeed the *attitude* toward diversity—rather than anything inherent in diversity itself—that creates the problem.

Bill Johntz, the founder of Project SEED in California, often spoke about the value of mathematics in overcoming the effects of cultural diversity. He insisted that mathematics is a great leveler, since it tends to be a new culture and a common culture, because of its newness, for all students.

Mathematics provides students with the opportunity to learn another "language," according to Johntz. His success over several decades in teaching traditionally low-performing students to perform at the highest levels in mathematics achievement certainly should cause us to pay attention to his analysis of the role of culture in the teaching of mathematics (Egan 1994; Barrett 1992; Beauchamp 1992; Carter 1976; Cummings 1978).

In fact, Johntz argued that the mathematics teacher more than any other teacher is at the forefront of demonstrations regarding the raising of both students' academic self-esteem and the students' intellectual potential, with self-esteem rising because of the phenomenal rise in student achievement. In fact, Johntz's work in mathematics was primarily aimed at raising students' self-esteem. The wonderful growth in both arithmetic and mathematics achievement was a byproduct of the decision to use this prestigious subject as the vehicle for the change in student image.

TECHNOLOGY AND MATHEMATICS TEACHING

Technology will prove of increasing value in mathematics instruction. The beautifully sophisticated software that is now available has the prospect of democratizing mathematics achievement, provided that the hardware and software are made accessible to all.

The availability of this software may even help to overcome some of the deficiencies in teacher preparation in mathematics. In fact, we have the unique tradition in the United States of teachers learning more about teaching methods than about the content of subjects for which they are responsible, especially in the area of mathematics.

There seems to be a correlation between teacher preparation in mathematics and assignments to classes based on socioeconomic status. Poor and cultural minority students have less access to teachers who are highly prepared in mathematics. The mathematics insights that students can achieve at very early ages tend to be far keener than the mathematics insights many of their teachers possess.

Naturally, if students are capable of high-level mathematics thinking and problem solving, teachers are also. But in view of past practice, a massive staff development program must be undertaken to rescue thousands of teachers from the consequences of previous, inappropriate staff development.

This matter is urgent. Many students have access to laptop computers and hand-held calculators while others go through a whole school career without seeing either. This is what Kozol (1991) meant by "savage inequalities." While these tools are important, it is also important for teachers who know how to use them appropriately to be made available to students if students are to gain the most from them.

The technology can be liberating, assisting the teacher in building an academic foundation from which student learning can be accelerated and extended. But the problem is student access to technology and to technologically sophisticated teachers (Hilliard 1985).

Let's make the point as precisely as possible. Without any special text materials, without any special computer technology, and without any special "minority" teacher training program, many teachers are able to help their students achieve the highest academic standards. In fact, there is research to show that often the formerly low-achieving cultural minority students can be taught so that they outperform the traditionally high-performing "middle class" students of any ethnic group. Does this happen because of specialized responses to cultural diversity and to the issues cited above?

The fact that high levels of mathematics achievement are attainable by all children without special technology is not at all an argument against the use of technology. On the contrary, computers, CD-ROM, and the information superhighway must be made available to all students.

Learning mathematics well must be a high priority. However, we must also guarantee that all students practice applying mathematics to technology and to and through the exceptional software that is available. We cannot tolerate a technology gap between the rich and the poor, between the minorities and the majority.

UNDERSTANDING LOW ACHIEVEMENT

High levels of mathematics achievement are attainable without reference to special education pedagogy, without mental measurement, and without cultural change in the schools. But this is not in any way an argument against special education, assessment, or cultural sensitivity and responsiveness; it is merely an effort to divide the problem of low achievement into components.

The basic division is between regular common pedagogy, which produces high student achievement in mathematics regardless of culture and socioeconomic background, and specialized pedagogy and services that enhance the effects of regular pedagogy of high quality. The implication here is that our primary problem is the inequitable distribution of good teaching and school services (Kozol 1991). We cannot begin to address diversity issues among students until we address diversity issues in the delivery of school services.

For some teachers, the surprising reality is that the low achievers change, not because of the invention of some culturally specific exotic new method, but because many of them are exposed to regular high-quality instruction, to which many of the traditionally low performers had not been previously exposed. It is hard for many educators and policymakers to see this. To do so means accepting the awful reality of savage inequities that permeate the schools.

Few of us are prepared to accept the reality of the magnitude and pervasiveness of these inequities. As a result, when students do not achieve, our first thought is to search for pathology in the child, the family, or the community. We seem unable to accept the diversity among us in levels of pedagogical expertise, levels of content knowledge, and levels of comfort with cultural and other forms of diversity. Therefore, we find it difficult to improve our knowledge and skills, because we do not see these things as problematic. We spend our time analyzing the wrong parts of the equation.

We must always remember that we work in an environment where powerful conflicts are in operation between and among groups of different kinds. Teachers and others engaged in the education of children are not at all immune to these conflicts. Consciously or unconsciously, many of us are actually partisans. Racial segregation and ideologies that assert that some racial or ethnic groups are intellectually superior to others are symptoms of these conflicts (Snyderman and Rothman 1988).

It takes effort and leadership to acknowledge and to overcome these forces. But the professional teacher who can overcome the politics of the environment is well positioned to learn those things about teaching and learning that will ensure his or her success with children from any group.

Anyin Palmer, principal of the Marcus Garvey School in Los Angeles, and his lead staff developer told me about the three simple criteria that they use to select teachers. Does the teacher love the children? Is the teacher patient with the children? Does the teacher love herself or himself? Teachers who meet Palmer's three simple criteria are then apprenticed to an experienced teacher.

Palmer is certainly entitled to speak and to be heard. The Marcus Garvey School begins algebra instruction in kindergarten. Its fifth graders are introduced routinely to beginning calculus. Its third graders have challenged and defeated sixth graders from the Los Angeles Unified School District's magnet school for the gifted in mathematics, even though it is a school in a low-income area of Los Angeles, populated totally by African Americans (Helmore 1986; Douglas 1981).

Few of the Garvey School teachers possessed a college degree! Their success was due to the right attitude and well-designed staff development. Of course, this is not an argument against degrees, certificates, or staff development—quite the opposite. It is *both* an argument for *valid* staff development and a challenge to the validity of much traditional staff development.

There are many highly effective staff development activities. Project SEED, which teaches higher mathematics to K–6 low-income minority cultural groups, has an excellent staff development process. So does Beginning School Mathematics, the New Zealand preschool program that teaches algebraic thinking to three-year-olds (for more information, contact Donald Miller, Oakland University, School of Education, Rochester, MI 48309). So does "Professor B," Everard Barrett, who teaches higher mathematics to young students. For example, most fifth graders in his class in Bedford-Stuyvesant, in New York City, passed the ninth-grade State of New York Regent's Examination (Barrett 1992; Cummings 1978).

Staff development is the problem! But just as it is true that young children are almost universally gifted with mathematics potential, so it is that teachers possess the talent to learn how to be successful with any child, regardless of his or her ethnic background. However, this success will occur only if there is a shift of emphasis in problem solving, from the diagnosis of "disabilities" in children to ensuring that teachers receive valid staff development.

STAFF DEVELOPMENT ISSUES

There are certain simple things about staff development in mathematics. First, the heaviest part of the staff development for new teachers or for unsuccessful older teachers should take place in real classrooms. It should be hands-on. Second, "master teachers" who have the responsibility for staff development *must themselves be able to* demonstrate *that they can raise the academic achievement of all children*. A primary reason for the failure of most staff development is our lack of quality control in selecting staff developers who are successful practitioners. Third, virtually all methods of instruction work; the big question is whether any method is used appropriately, that is, again, quality control.

Teachers can be successful using relatively unsophisticated methods, if they persist, if they know the subject, and if they meet Anyin Palmer's criteria. On the other hand, if teachers are willing to learn some of the more sophisticated methods, they can learn to work smarter rather than harder. What is important is that we do indeed know what must be done in order to be successful with all children. The question is, as I have asked before, "Do we have the will?"

The good news is that, where mathematical excellence for cultural minority students is concerned, *there is no problem other than the problem of whether we will embrace the solutions that have already been manifest.* Therefore, the task before us is clear. We do not have a discovery problem. We have a dissemination problem. We have an accountability problem.

SUMMARY

Deservedly or not, mathematics achievement can be the basis for prestige. One's academic and intellectual self-image is often tied to mathematics achievement. One's professional self-esteem is also tied to identification as a mathematics teacher. One's national identity is similarly affected.

Therefore, whether or not mathematics is as difficult as it is reputed to be—or whether or not the esteem that mathematics achievers feel and are accorded is earned—there is tradition. But given the power of teaching, this achievement is accessible to all.

Mathematics learning offers thought and language of precision, patterns, rhythms, harmonies, and beauty. All children deserve to have access to this knowledge. It is a gift of the eternal. A mathematics teacher ought to be a teacher with a mission. To show that there is no learning problem in mathematics is to challenge every academic area to confront the awesome truth.

Human learners are awesome. Human genius is universal, independent of "race," socioeconomic status, language, or gender. Let us release the genius of all the children.

. . . .

BIBLIOGRAPHY

Backler, Alan, and Sybil Eakin, eds. *Every Child Can Succeed: Readings for School Improvement.* Bloomington, Ind.: Agency for Instructional Technology, 1993.

Barrett, Everard. "Teaching Mathematics through Context: Unleashing the Power of the Contextual Learner." In *Nurturing At-Risk Youth in Math and Science: Curriculum and Teaching Considerations,* edited by Randolf Tobias, pp. 49–78. Bloomington, Ind.: National Educational Service, 1992.

Beauchamp, Lane. "Where Every Child Is Gifted: Fairbanks Elementary, an Accelerated School in Springfield, Mo., Challenges Students from Troubled Homes by Focusing on Their Strengths." *America's Agenda* (Spring 1992): 32–4.

Carmichael, J. W., Jr., and John P. Sevenair. "Preparing Minorities for Science Careers: What's the Secret of a Small University's Big Success?" *Issues in Science and Technology* (Spring 1991): 55–60.

Carter, Sylvia. "Roosevelt Program Adds Up to Success." *Newsday*, 23 June 1976, pp. 4A, 5A.

"The Common Language." *Newsweek*, 4 May 1970.

Cummings, Judith. "Math Method Deducts Rote and Adds to Pupils' Scores." *New York Times*, 9 June 1978.

Diop, Cheikh Anta. *Civilization or Barbarism: An Authentic Anthropology.* Brooklyn, N.Y.: Lawrence Hill, 1991.

Douglas, Pamela. "Private Schools: By, and for, Blacks. L.A. Students Use College Arithmetic Text in 3rd Grade." *Atlanta Journal and Constitution*, 4 October 1981, p. 28-A.

Egan, Timothy. "Accelerating Poor Achievers." *New York Times*, Education Life section, 2 August 1992.

Edmonds, Ron R. "Some Schools Work and More Can." *Social Policy* (March/April 1979): 28–32.

Feuerstein, Reuven. *The Dynamic Assessment of Retarded Performers: The Learning Potential Assessment Device.* Baltimore, Md.: University Park Press, 1979.

_____. *Instrumental Enrichment.* Baltimore, Md.: University Park Press, 1980.

Hall, E. T. *Beyond Culture.* New York: Anchor, 1977.

Helmore, Kristin. "When Great Expectations Pay Off." *Christian Science Monitor*, November 1986.

"Higher Math in Lower School." *Yale Alumni Magazine* 34 (9) (July 1971): 24–5.

Hilliard, Asa G. III. "Behavioral Style, Culture, and Teaching and Learning." *Journal of Negro Education* 61 (3) (1992): 1–8.

_____. "Do We Have the Will to Educate All Children?" *Educational Leadership* (September 1991): 31–6.

_____. "Sociopolitical Implications of Competency Testing." In *Minimum Competency Education: Issues, Methodology, and Policy for Local School Systems*, edited by John H. Neel and Shirley W. Goldwasser. Atlanta: Georgia State University, 1982.

_____. "Teachers and Cultural Styles in a Pluralistic Society." *NEA Today* 7 (6) (1989): 65–9.

_____. "What Good Is This Thing Called Intelligence and Why Bother to Measure It?" Paper presented to American Educational Research Association Symposium:

"Whatever Happened to the Measurement of Intelligence?" at AERA annual meeting, New Orleans, April 1994.

Jones, J. Arthur. "Look at Math Teachers, Not `Black English.' " *Essays and Policy Studies*. Washington, D.C.: Institute for Independent Education, 1987.

Kozol, Jonathan. *Savage Inequalities: Children in America's Schools.* New York: Crown, 1991.

Lidz, C. S., ed. *Dynamic Assessment: An Interactional Approach to Evaluating Learning Potential.* New York: Guilford Press, 1987.

_____. *Practitioner's Guide to Dynamic Assessment.* New York: Guilford Press, 1991.

Mathews, Jay. *Escalante: The Best Teacher in America.* New York: Henry Holt & Co., 1988.

_____. "Escalante Still Stands and Delivers: A Great Math Teacher Tackles New Problems." *Newsweek,* 20 July 1992, pp. 58–9.

Orr, Eleanor W. *Twice As Less: Black English and the Performance of Black Students in Mathematics and Science.* New York: W. W. Norton & Co., 1987.

"Seed: New View of Learning." *National Observer,* 19 August 1968.

Sizemore, Barbara. "The Algebra of African-American Achievement." In *Effective Schools: Critical Issues in the Education of Black Children,* edited by Percy Bates and Ted Wilson, pp. 123–49. Washington, D.C.: National Alliance of Black School Educators, 1988.

Sizemore, Barbara, Carlos Brosard, and Berny Harigan. *An Abashing Anomaly: The High Achieving Predominately Black Elementary Schools.* Pittsburgh: University of Pittsburgh Press, 1982.

Skrtic, Thomas M. "The Special Education Paradox: Equity as the Way to Excellence." *Harvard Educational Review* 61 (2) (1991): 148–206.

Snyderman, Mark, and Stanley Rothman. *The I.Q. Controversy: The Media and Public Policy.* New Brunswick, N.J.: Transaction Publishers, 1988.

Spindler, George, ed. *Education and Cultural Process: Anthropological Approaches.* Prospect Heights, Ill.: Waveland Press, 1987.

Walters, Laurel. "Students Take the Fast Track: 'Accelerated Learning' Seeks to Engage At-Risk Pupils, Not Relegate Them to Remedial Classes." *Christian Science Monitor,* 10 December 1990.

Weinberg, Meyer R. *A Chance to Learn.* Cambridge: Cambridge University Press, 1977.

Wiggins, Grant P. *Assessing Student Performance: Exploring the Purpose and Limits of Testing.* San Francisco: Jossey-Bass Publishers, 1993.

Zaslavsky, Claudia. "'Africa Counts' and Ethnomathematics." *For the Learning of Mathematics* 14 (2) (June 1994): 3–8.

_____. "Mathematics in Africa: Explicit and Implicit." In *Companion Encyclopedia of the History and Philosophy of the Mathematical Sciences,* edited by I. Grattan-Guinness, pp. 85–91. New York: Routledge, Chapman & Hall, 1994.

Trends in Mathematics Achievement for Young Men and Women

Nancy Burton

*E*VERYBODY *knows that boys are better at math than girls.* Or is that an old uncle's tale that should have been debunked years ago? This paper will review the evidence that young women don't perform as well as young men in math. In brief, girls have traditionally earned higher grades in math than boys, just as they have traditionally earned higher average math test scores up until junior high school.

In the last twenty years, young women have been taking more and more advanced math courses, and they now earn the same average math scores as young men through high school. However, the data show that among higher-scoring students, on more advanced content, and on unusual problem types, young men still score higher than young women. Some of the possible causes, including bias in test scores, bias in test items, bias in grades, and bias in the way girls are treated at home and in school, are critically reviewed.

WHAT IS KNOWN

What Was Known Twenty Years Ago

When researchers started looking systematically at gender differences in math, they found that matters were not quite so simple as the folk beliefs

· · · · · · · · · · · · · ·

Nancy Burton is director of development, analysis, and research for the SAT at Educational Testing Service. She has spent two decades working with large-scale assessments, including the NAEP, GRE, and SAT. Her research focuses on issues of test validity and fairness for gender and ethnic groups.

"everybody knew." Contemporary research in gender differences can be dated from the publication, in 1974, of Maccoby and Jacklin's definitive review and interpretation of the research literature. Maccoby, Jacklin, and the many researchers following them found that, for example, girls almost always got better average math grades than boys. When researchers looked at test scores, they found that up until about junior high school, girls on average scored better than boys, and that thereafter girls' average math scores began to fall below boys'.

Women have traditionally scored much lower than men on the Graduate Record Exam (GRE) quantitative test. The first National Assessment of Educational Progress (NAEP) in math, in 1971–72, showed that the gender differences were substantially larger for young adults than for high school students. In general, the gender differences widened as students got older. The crossover from young women's scores being above young men's scores to being below occurred just about at puberty.

Looking at math content, young women of all ages tended to score better on arithmetic, about the same on algebra, and sometimes lower on geometry, especially where visualization and manipulation of figures were required. It was hard to compare women's and men's performance on higher-level math, because most young women stopped taking math as soon as they were allowed the choice.

Looking at the entire range of performance, rather than just the average performance, researchers noted that young women obtained the full range of scores that young men did. In other words, even though there *was* often a difference in average scores, the distributions of scores almost completely overlapped, and girls displayed the entire range of mathematical talent. But girls, their parents, their counselors, and even—though of all people they surely knew better—their math teachers too often behaved as if every single girl were worse at math than every single boy.

Researchers had not even completed filling in this picture when it began to be clear that the numbers were changing.

What Is Known Today

Meta-analysis of mathematics test scores

Researchers had begun to use "meta-analysis," a systematic, statistical way to combine results across many research studies, first proposed by Glass in 1976. This is a particularly good technique for describing overall patterns and for illuminating small differences that are difficult to detect in any single study.

Some of the most interesting recent meta-analysis work in math has been done by Feingold (1988, 1993); Friedman (1989); Hedges and Olkin (1985); Hyde, Fennema, and Lamon (1990); Linn and Petersen (1985); and Linn (1986). These researchers began to note that the more recent studies showed smaller average gender differences.

By the late 1980s, Feingold (1988) stated, "cognitive gender differences are disappearing." Although that announcement was somewhat premature, it was based on a very substantial finding. Using data from norms studies by various test publishers, Feingold was able to show that average reading and math gender differences had declined to virtually zero for representative national samples of high school students. For example, the Preliminary Scholastic Assessment Test (PSAT/NMSQT) was given to representative samples of high school juniors (in addition to the group of college-bound students who elect to take the test each October) in 1960, 1966, 1974, and 1983. Over those years, the average gender difference in math declined from 3.3 points (about a third of a standard deviation) to 1.4 points (12 percent of a standard deviation) in the representative norms samples.

Trends in course-taking

This trend in scores needs to be evaluated in light of contemporaneous trends in course-taking. During the years that the average female-male math score difference declined, the gender differences in math and science course-taking also declined. By 1993, for example, the female/male split in senior SAT takers was as shown in table 1.

• •

Table 1. Female/Male Split in Senior SAT Takers

Course	Percent Female/Male
Algebra	54/46
Geometry	53/47
Trigonometry	52/48
Precalculus	51/49
Calculus	47/53

• •

In other words, 54 percent of the students who reported taking algebra were women and 46 percent were men. Note that since 53 percent of SAT takers are women, a strictly proportional rate of participation by sex would be 53/47.

In the more quantitative science courses, the gender breakdown was as shown in table 2.

· ·

Table 2. Gender Breakdown in Quantitative Science Courses

Course	Percent Female/Male
Biology	54/46
Chemistry	53/47
Physics	47/53

(College-Bound Seniors: 1993 Profile)

· ·

The higher-level courses—calculus in math, physics in science—show lower participation by women than men. However, trends in participation are still changing rapidly. Only five years earlier, in 1988, the calculus gender breakdown was 44/56, and the physics gender breakdown was 43/57.

In short, the increase in women's scores relative to men's is surely related to the increase in women's course-taking.

Questions That Remain

Researchers have continued to uncover complications and exceptions to the overall trends. One of the most active areas of concern has been the small number of women receiving very high math scores. This has generated media attention because of the direct effect of math scores on the chances of receiving prestigious scholarships such as the New York Regents, the National Merit, and the New Jersey Governor's, for example.

Gender differences in top scores

A picture is beginning to emerge. Men's math performance tends to exceed women's

- on more advanced content,
- on more unfamiliar or unusual problems,
- on problems presented in unusual formats or contexts,
- in more selective groups, and
- in older groups.

These categories are difficult to disentangle. Older groups may be taking tests on more advanced content and be more selective, too: for example, comparing GRE takers to SAT takers to high school standardized test–takers. Similarly, difficult content is more likely to be presented in unusual problems or formats.

This kind of confounding has happened before. For example, there was a long debate about whether women's declining scores after puberty had to do with hormones or course-taking. Now that women are taking more math in high school, and their test scores are equal to or better than men's until the end of high school, that particular question seems to be answered.

Another kind of confounding has to do with continuing ripples from earlier trends. For example, older women may still be scoring lower today because they grew up in an earlier era. If one dates the beginning of the current emphasis on encouraging girls to take math back to the issuance of the Women in Math grants by the Office of Education in the mid-1970s and assumes that any national program will take at least five years to begin having effects at the local level, then the first women to have gone through school being encouraged to take math are just now beginning to appear in graduate school.

Despite the fact that the current evidence cannot be definitive, however, the general pattern of increasing gender differences on increasingly difficult content does seem to make sense. It is consistent with the statistics that show today's gender differences in taking advanced courses such as calculus and physics.

It also fits in with psychological theories about how learners move from novice to expert status in a field. That is, young women tend to do best on math problems that are similar to problems presented in math class, particularly on problems in which the algorithm is supplied. In this they behave like novice learners, who are not yet comfortable using their skills in unfamiliar situations.

Gender differences in beliefs about mathematics

The results of one of the more recent National Assessments of Math (NAEP 1988) showed a potentially influential gender difference for nine-year-olds. On most of the problems, the girls performed as well or better than the boys. On background questions, the girls were just as likely as the boys to say they liked math and were good at math. However, the girls were much less likely to say that math would be useful to them in the future.

If one treats information that one expects to use over and over again differently from information that one will never need after the semester test, this belief of nine-year-old girls could have important long-term effects. Age nine is likely to be crucial in allocating one's cognitive territory. If math gets relegated to some mental attic, the chances of developing a rich and pervasive understanding of math, available to be applied to the varied problems of daily life, must be seriously reduced.

Efforts to encourage young women to stay in math classes in high school, and even those to encourage girls to get into algebra as early as possible, cannot overcome such early decisions.

Even these somewhat late efforts have had positive effects. Now young women tend to stay in math courses through high school, and their average scores have risen to reflect that course-taking. Today's young women have not prematurely closed down their options for careers in science and technology. And, simply by persevering and continuing to succeed, more young women have learned to question the still-strong cultural assumption that math and science careers are for men. However, there is still a long way to go.

Gender differences after high school

In 1993, the same young women in the SAT population (who had taken nearly equal high school mathematics and science courses with young men in high school) showed more sex-stereotyped plans for college. When asked what their plans for college majors were, the class of 1993 responded as in table 3.

· ·

Table 3. Female SAT population's plans for college majors

Major	Percent Female/Male
Traditional Fields for Men	
Mathematics	46/54
Physical Sciences	35/65
Computer Science	33/67
Engineering	19/81
Traditional Fields for Women	
Social Science/History	67/33
Health & Allied Services.	68/32
Education	77/23

(College-Bound Seniors: 1993 Profile)

· ·

Astin's *American Freshman: National Norms for Fall 1993,* based on a completely different national sample of freshmen in all institutions, showed very consistent results to the SAT *1993 College-Bound Seniors* for majors. Astin also asked the freshmen to name their probable career. Interestingly, the freshmen's probable career choices also tracked very closely with their intended majors (see table 4).

The gender breakdown of SAT takers reporting a calculus course in high school (46/54) is equal to the gender split of fall freshmen enrollees who intend to major in math (46/54) and the gender split of students who are considering a career in the general area of scientific research (46/54). (Remember that women are 53 percent of all SAT takers and 54 percent of all college freshmen, so women's scientific aspirations are somewhat lower than these charts make them appear.)

• •

Table 4. College Freshmen's Probable Career Choices

Career	Percent[1] Female/Male
Traditional Fields for Men	
Scientific Research	46/54
Computer Science	33/67
Engineering	18/82
Traditional Fields for Women	
Nursing/Physical Therapy	79/21
Education	74/26

(The American Freshman: Fall 1993)

[1] *Astin's tables simply listed the percent of women choosing a certain major or career separately from the percent of men. In order to convert that to the percent of the total group choosing a given career, the percents were weighted by the proportion of women to men in the total weighted sample (54/46, Table A3, page 109).*

• •

These results should be interpreted cautiously, since students' aspirations at this point are not at all fixed and the seriousness of their answers to questionnaires of this sort is open to question. Nevertheless, one cannot enter a career if one never considers it. Particularly in scientific and technical areas, where the progression of required courses tends to be inflexible, dropping out of science and math as a freshman can seriously delay any future move back onto the scientific track.

Efforts to encourage women's participation in math and science, while helpful, so far have only delayed the stage when young women tend to leave the track. They used to drop out in high school, as soon as they were given some choice of electives. Now they wait until college.

WHAT CAN BE DONE

Advice from Test Critics

Some critics have suggested that tests showing gender differences should be changed, abandoned, or discounted (for example, Rosser [1989]). There seem to be two reasons for this advice. First, it is argued that young women tend to get better grades in math courses in high school and, though to a lesser extent, in college, so the math tests showing better performance for males than for females must be biased against women.

Second, some critics argue that young women tend to be very responsive to external evaluations. They say that lower test scores have an especially negative effect on women's participation, that young men are more likely to persist despite external discouragement while young women are more likely to quit. Both of these arguments deserve serious consideration.

Are the tests biased?

The observation that young women tend to have better grades in mathematics classes even when they do not score better on standardized mathematics tests is well established. These data are almost always based on predictive validity studies, in which objective variables available at the time a decision is being made are used to predict a measure of future success. The studies most commonly done with the SAT, predictive studies to help in the selection of a freshman college class, typically use these predictors:

- Test scores believed to provide information on skills generally useful in college, such as the SAT V and M scores or the ACT composite score.
- High school grade point average, which is considered to be a broader measure of academic background and which includes such important factors as motivation and study skills.

Test scores have the disadvantage of being quite narrowly focused (after all, they are based on two to three hours of testing), but they have the compensating advantage of being common across all applicants.

Grade averages are based on three or four years of work but have other disadvantages. Different patterns of course-taking are all treated as equal. Different grading standards across teachers, departments, and schools are all treated as equal. Grades are subjective, with some teachers caring more than others about poor attendance or penmanship or disruptive behavior, or.... Finally, teachers are subjected to pressure by parents, coaches, or principals to modify grades, and some do.

The SAT was developed in 1926 to help compensate for some of the disadvantages of high school grade averages. It was meant to supplement high school grades and to be an alternative means of access for students who either attended a high school with a poor academic reputation or students who had not done well in high school. In fact, the SAT and high school grade point average do complement each other. Together, grades and test scores do a moderately good job of predicting success in college as measured by freshman grade point average.

Particularly in light of charges of racial and gender bias in tests, a great deal of research has been done on how tests are used in college admission. Research has focused on all aspects of the admission process, including the existing variables used as predictors, the variables used as criteria of success in college, the admission decision process itself, and, finally, on a search for other variables to use as predictors.

Research on new predictors

Research attempting to find variables other than high school grades and standardized tests have been largely unsuccessful. The academic variables studied all tend to be very highly correlated, so even measures that *sound* as if they are measuring something new really just duplicate the information provided by existing test scores. Nonacademic variables such as motivation tend to be too easy to fake or ethically objectionable to use in making admission decisions. The most promising work, by Willingham (1985), identified "successful persistence" in extracurricular activities as a generally useful and independent predictor of success. Colleges, however, have not rushed to use it, perhaps because such activities are relatively difficult to quantify and to verify.

Research on the decision process

Throughout the 1970s, much very good statistical work was done to improve the selection process using game-theory and decision-theory mathematical models. The research was successful but it has had little effect on admission practices, probably because the models are both hard to understand and cumbersome to implement.

Research on admissions tests

Similarly, much research was done and continues to be done on questions of item bias and test bias. This research led to some improvement in test development and use, but the improvements so far have been small. Thousands of studies exist showing that in general, admissions tests predict college grades about as well for men, women, majority, and minority students.

Hundreds of studies now exist examining the fairness of individual test items. Some progress has been made in identifying item contexts that may favor one group or another. For example, math problems in a sports or mechanical context tend to be more difficult for women. This tendency is very slight. Math items show very little "differential item functioning" (DIF)—perhaps one item in a hundred will be systematically more difficult for women or black students or Hispanic students or Asian students.

With sports items, the probability is slightly higher—say, three items in a hundred. Educational Testing Service (ETS) continues to try to identify content differences among items, but in general relies on statistical tests to find suspect items, given the rarity of bias even in content categories known to be problematic. The SAT program does a DIF analysis as part of the pretest analyses, removing any suspect items before they ever appear on scored SAT editions.

Because the DIF statistic is not particularly reliable, many of the items being dropped have no bias; however, since item bias is so difficult to recognize, it is safest simply not to use items flagged by the DIF analysis unless they are required by the test specifications.

Recently, two promising streams of research have begun.

Research on performance assessment

The work that has begun in the areas of performance assessment and portfolio assessment is very promising in that it should allow students to present their own unique accomplishments. This should be beneficial for those minority and women students who are not following the expected path. However, the problems of measuring complex performances fairly are formidable: It may be years before the effects of this new emphasis in measurement will be known.

Research on freshman grade average

The second promising area of research has already had substantial results. In the 1980s, a number of researchers from widely separated institutions began to wonder about the quality of freshman grade average as a criterion of success in college. Some of the most important work in this area has been done by Elliott and Strenta (1988) at Dartmouth; McCornack and McLeod (1988) in California; and Ramist, Lewis, and McCamley (1990), Clark and Grandy (1984), and Gamache and Novick (1985) at ETS and the American College Testing Program (ACT).

Work began first with attempts to adjust for differences in grading standards across divisions of the university. However, it was not until the late 1980s that researchers began to look at performance within courses. The results of this work were quite compelling. When high school grades and test scores are used to predict grades within classes, gender differences in prediction tend to disappear.

The reason is quite simple. As shown above, young men are much more likely than young women to take courses in engineering and physical science. Engineering and physical science instructors tend to give systematically lower grades than instructors in other disciplines. What is meant by "systematically lower" is that the grades are lower than one would expect given the overall quality of the students' backgrounds.

Science instructors give relatively low grades to students with good high school preparation in science, good test scores, and good grades in other college courses. As a result young men, who take more science than average, get lower grades than average, while young women, who tend to stay out of science, get higher grades than average.

This has nothing to do with the true ability of young men and women or with the true difficulty of science courses compared to others. It's just what must happen arithmetically when one group of instructors uses a different standard. This artifact of the science culture explains almost all the difference in prediction that has been noticed for men and women.

This is not the whole story of grades. In addition to these systemic differences in grading standards, there are some individual differences in learning skills and in attitudes that help the average young woman get slightly higher grades than the average young man, all else being equal. The fact that both high school and college grades contain information about academic competence *not* captured by tests emphasizes the importance of not relying upon test scores alone whenever possible.

Do lower test scores discourage young women from entering mathematics and science?

The second claim by critics is that young women are more affected by test scores than young men, so that disproportionate numbers of young women choose not to take math and science courses in which they could perform perfectly well. This point remains whether or not the test scores of young women are accurate. Critics argue that even if the low scores are correct, the benefit of the information provided is overshadowed by the cost to science of the loss of female participants. This is a serious argument, since test publishers *should* be held responsible for the unintended consequences of test use.

However, suppose that this allegation had been established in fact and women's mathematics and science scores had been suppressed. Set aside for the moment the difficulty of fair and useful college admission based only on verbal skills in this technological era. The research and policy discussion of gender differences that has been growing since the early 1970s could not have occurred. The successful efforts to encourage women to persist in science and mathematics courses through high school would not have occurred. The question raised in this paper about the influence of beliefs about mathematics formed at a very young age could not arise.

In the absence of empirical information, would the folk belief that math and science are not for girls eventually fade? Thirty years ago, before the current interest in gender research, the folk belief was actually overgeneralized. The research of the last thirty years has tended to narrow the perceived size of gender differences and to limit the content areas where gender differences have been found. I certainly believe that suppressing information is no way to close those gaps in young women's mathematics education that still exist. I believe that it is better to acknowledge the test score information once it is shown to be correct and to look for a way to improve women's math scores through improved education. In the long run it is unwise to ignore or suppress information.

. . . .

BIBLIOGRAPHY

Astin, A. W., W. S. Korn, and E. R. Riggs. *The American Freshman: National Norms for Fall 1993*. Los Angeles: Higher Education Research Institute, UCLA, 1993.

Clark, M. J., and Jerilee Grandy. *Sex Differences in the Academic Performance of Scholastic Aptitude Test Takers*. CB Report No. 84-8; ETS RR 84-43. New York: College Entrance Examination Board, 1984.

College Board. *College-Bound Seniors: 1993 Profile*. New York: College Board, 1993.

_____. *College-Bound Seniors: 1988 Profile*. New York: College Board, 1988.

Elliot, Rogers, and A. C. Strenta. "Effects of Improving the Reliability of the GPA on Prediction Generally and on Comparative Predictions for Gender and Race Particularly." *Journal of Educational Measurement* 25 (4) (1988): 333–47.

Feingold, Alan. "Cognitive Gender Differences Are Disappearing." *American Psychologist* 43 (2) (1993): 95–103.

_____. "Joint Effects of Gender Differences in Central Tendency and Gender Differences in Variability." *Review of Educational Research* 63 (1) (1993): 106–9.

Friedman, Lynn. "Mathematics and the Gender Gap: A Meta-analysis of Recent Studies on Sex Differences in Mathematical Tasks." *Review of Educational Research* 59 (2) (1989): 185–213.

Gamache, L. M., and M. R. Novick. "Choice of Variables and Gender Differentiated Prediction within Selected Academic Programs." *Journal of Educational Measurement* 22 (1) (1985): 53–70.

Glass, G. V. "Primary, Secondary, and Meta-analysis of Research." *Educational Researcher* 5 (10) (1976): 3–8.

Hedges, L. V., and B. J. Becker. "Statistical Methods in the Meta-analysis of Research on Gender Differences." In *The Psychology of Gender: Advances through Meta-Analysis*, edited by J. S. Hyde and M. C. Linn, pp. 14–50. Baltimore: Johns Hopkins University Press, 1986.

Hedges, L. V., and Ingram Olkin. *Statistical Methods for Meta-Analysis*. Orlando, Fla.: Academic Press, 1985.

Hyde, J. S., Elizabeth Fennema, and S. J. Lamon. "Gender Differences in Mathematics Performance: A Meta-Analysis." *Psychological Bulletin* 107 (2) (1990): 139–55.

Linn, M. C. "Meta-analysis of Studies of Gender Differences: Implications and Future Directions." In *The Psychology of Gender: Advances through Meta-Analysis*, edited by J. S. Hyde and M. C. Linn, pp. 210–31. Baltimore: Johns Hopkins University Press, 1986.

Linn, M. C., and A. C. Petersen. "A Meta-analysis of Gender Differences in Spatial Ability: Implications for Mathematics and Science Achievement." In *The Psychology of Gender: Advances through Meta-Analysis*, edited by J. S. Hyde and M. C. Linn, pp. 67–101. Baltimore: Johns Hopkins University Press, 1986.

Maccoby, E. E., and C. N. Jacklin. *The Psychology of Sex Differences*. Stanford, Calif.: Stanford University Press, 1974.

McCornack, R. L., and M. M. McLeod. "Gender Bias in the Prediction of College Course Performance." *Journal of Educational Measurement* 25 (4) (1988): 321–31.

National Assessment of Educational Progress. *The Mathematics Report Card*. Princeton, N.J.: Educational Testing Service, 1988.

Ramist, Leonard, Charles Lewis, and Laura McCamley. "Implications of Using Freshman GPA as the Criterion for the Predictive Validity of the SAT." In *Predicting College Grades: An Analysis of Institutional Trends over Two Decades*, edited by W. W. Willingham, Charles Lewis, Rick Morgan, and Leonard Ramist. Princeton, N.J.: Educational Testing Service, 1990.

Rosser, Phyllis. *The SAT Gender Gap: Identifying the Causes*. Washington, D.C.: Center for Women Policy Studies, 1989.

Willingham, W. W. *Success in College: The Role of Personal Qualities and Academic Ability*. New York: College Entrance Examination Board, 1985.

TEACHERS AND TEACHING . . .

A progressive image of teachers and the teaching profession is presented in this section. All stages of development, from recruitment and initial preparation to full status as professionals to maintaining commitment and professional expertise, are discussed herein. The decisions that affect teachers and the teaching environment in schools with increasingly diverse student populations are central to this issue.

Teaching for conceptual development, understanding, and student empowerment requires a very different instructional style from that of teaching for the mastery of technical skills. The first paper vividly portrays mathematics teachers pursuing their craft on a daily basis, demonstrating how school mathematics can and should be taught. Filling and refilling the pipeline that supplies future generations of classroom professionals must also be considered. The second paper calls for reforming college and university programs so that beginning teachers are properly prepared to start their journey toward a long and fulfilling career. The recruitment and retention of highly qualified mathematics teachers, particularly those from racial and ethnic minorities, is a prime concern. As part of the solution to this dilemma, trends that contribute to the shortage of minority teacher-leaders, particularly in schools and communities where they are most sorely needed, are analyzed in the third paper. The vision of mathematics teachers becoming reflective, lifelong learners of their craft through professional development programs is advanced in the fourth paper. The fifth paper concludes this section with an example of how this vision may be translated into a national reality.

The challenges presented in these papers are clear and multifaceted. However, the essential role the mathematics teacher plays in the lives of future generations makes these challenges worthy of society's sustained attention.

—Charles A. (Andy) Reeves

How Should Mathematics Be Taught?

Glenda Lappan & Diane Briars

> *Grandma, come see my new computer program.... See, you turn it on like this ... oops, we gotta reboot ... then, select the program.... See, you try it! ... David, how's kindergarten? What are you learning? ... We learned the "eency, weency spider climbed up the water spout," then we practiced writing 3s.*
> *(Robert Coldiron 1994)*

TECHNOLOGY has dramatically changed our lives, making the world smaller through high-speed communication and changing the knowledge needed for the workplace and full participation in society. And the rate of change is increasing. Clearly, the world in which our current students will live and work will be very different from the one in which they are being schooled. The National Council of Teachers of Mathematics' three *Standards* documents—*Curriculum and Evaluation Standards for School Mathematics* (1989), *Professional Standards for Teaching Mathematics* (1991), and *Assessment Standards for School Mathematics* (forthcoming)—present a vision of mathematics education to prepare students for the twenty-first century. In this chapter, we look at the vision of mathematics teaching portrayed in these documents.

Glenda Lappan is a professor in the Department of Mathematics at Michigan State University. Her research and development interests are in learning and teaching at the middle and high school levels. She is currently codirector of the Middle Grades Mathematics Project, an NSF-funded curriculum project for grades 6–8.

Diane Briars is a mathematics specialist for the Pittsburgh Public Schools and currently a director of the NCTM. She was cochair of the Purposes Working Group for the NCTM Assessment Standards for School Mathematics. She is a former middle school mathematics teacher.

IMPLICATIONS OF CURRICULAR GOALS FOR INSTRUCTION AND ASSESSMENT

A key element of the reform vision is change in the mathematics curriculum. The *Curriculum and Evaluation Standards* calls for change both in the content of the mathematics curriculum and the overall goals for students. It calls for a curriculum that recognizes the dramatic impact of technology on what mathematics is important and what mathematics is possible. It also calls for new goals for students that include learning to value mathematics, to communicate and reason mathematically, and to become mathematical problem solvers who are confident in their ability to do mathematics. Detailed discussions about the reform in content are provided elsewhere in this volume (Steen and Forman, Seeley, and Mills).

This change in curriculum and in overall goals for students requires a change in instruction. Teaching to develop problem solving, reasoning, and communication skills requires a very different style of instruction from teaching technical skills such as computational algorithms and procedures. Reform is about teaching for understanding and teaching to promote students' confidence in themselves as mathematics learners and doers, rather than teaching students to be proficient at executing standard procedures.

To support this change in mathematics and teaching practice, assessment techniques and goals must change. The *Assessment Standards* (forthcoming) portrays a vision of assessment as an integral part of instruction, embedded in the flow of classroom activity, to provide both the student and the teacher with evidence of the growth of understanding, progress, skill, and disposition. Assessment activities are opportunities for further learning, for students' self-assessment, and for teacher instructional decision making. If this is the vision of assessment in support of standards-based reform, then the importance of the kinds of assessment tasks that teachers use is underscored. The alignment of curriculum, instruction, and assessment becomes the key component of changing classroom paradigms.

CHANGING CLASSROOM PARADIGMS

Another major influence on our vision of mathematics teaching is our view of learning. Research during the past decade on children's thinking is changing our conception of what learning is and how it occurs. This in turn is changing our perception of the teacher's role in promoting learning. In this section, we examine the implications of this emerging view of human learning for teaching mathematics.

Understanding Student Understanding

A natural question to ask in conjunction with "How should mathematics be taught?" is "How do students learn?" Research from cognitive science and mathematics education is producing a growing body of evidence that supports a view of learning as "the process of an individual mind making meaning from the materials of its experience" (Knoblauch and Brannon 1984). This viewpoint, referred to as *constructivism* (Morrow 1991), recognizes that students actively build their own mathematical knowledge from their experiences and rely on their peers, tutors, teachers, and themselves for feedback. This approach holds that:

- Learning is contextual: what students learn is fundamentally connected with how they learn it.
- Learning occurs best through dialogue, discussion, and interaction.
- Learners must be actively involved in the process.
- A variety of models must be used to meet the need of all learners (e.g., working individually, in pairs, and in cooperative groups).
- Learners benefit from reviewing, critiquing, and revising one another's work.
- The discovery approach to learning engages the learner and allows him or her to own the process.
- The discovery approach should lead to creating the language, syntax, and structure of mathematics.

This view of learning as invention or creation and abstraction has clear implications for instruction. Most notably, it *values students' thinking.* Students are seen as "thinkers with emerging theories about the world" (Brooks and Brooks 1994) rather than passive recipients of information. This calls for a shift from traditional teacher-centered instruction, in which "teaching is telling," to student-centered instruction, in which students are actively engaged in making sense of situations. The teacher's role shifts from dispensing information to selecting tasks, creating an environment, and orchestrating classroom discourse to foster student understanding. (See the Student chapters in this volume for more detailed descriptions of how students think and learn.)

This emphasis on active learning is really nothing new. The same sentiment is expressed in the ancient Chinese proverb: "Tell me, I forget. Show me, I remember. Involve me, I understand." Similarly, the use of concrete models and manipulative materials to help students make sense of abstract mathematical concepts is an idea that dates back to at least the early 1900s with Maria Montessori, if not earlier. What is new, however, is our increased understanding of the learning process and how these techniques can be used to enhance learning.

Contextualized Learning

One of the most interesting advances in cognitive science in the past two decades is a better understanding of the role of context in learning. The work of Carraher, Carraher, and Schliemann (1985) and others has made us consider the amazing mathematical proficiency children selling goods on the street have at the same time that these children are unable to perform in "school" math. The real context gives the activity meaning. When we consider the context-free study of arithmetic and the sanitized, simplified, often contrived contexts of algebra, it becomes clear that the power of situations has seldom been available for students of mathematics.

Brown, Collins, and Duguid (1989) talk about the relationship between what is learned and the activity in which that learning takes place.

> The activity in which knowledge is developed and deployed ... is not separate from or ancillary to learning and cognition. Nor is it neutral. Rather, it is an integral part of what is learned. Situations might be said to coproduce knowledge through activity. Learning and cognition ... are fundamentally situated.

A compelling example of this situated cognition happened in a seventh-grade classroom a couple of years ago. A student came bursting into the room the day after the Michigan Council of Teachers of Mathematics Middle School Exam had been given to all the middle school students in the school. "Dr. Lappan, the basketball problem was on the exam!" shouted the student. On further investigation, the basketball problem was not on the exam, but a problem that had the same mathematical structure and yielded to the same problem-solving technique was on the exam. For this student the context was a powerful organizer of what he had learned in struggling with the basketball problem.

To develop more productive notions about mathematics, students must have opportunities to actually be involved in doing mathematics in contexts—to explore interesting mathematical situations, to look for patterns, to make conjectures, to look for evidence to support their conjectures, and to make logical arguments for their conjectures. The situations in which these activities are embedded become part of what is learned and how that learning is remembered and recalled.

However, teachers have a mathematical agenda. Hence in choosing situations, not only does the teacher make judgments about the relevance of the situation to his or her students, a teacher judges how well the situation or task represents the embedded concepts and procedures that the teacher is trying to teach, how likely the students are to bump into the mathematics in the course of investigating the problem, and what skill development the task will or can support. The teacher asks, "With what mathematics does the situation and the task surround the student?"

The view of learning as making sense of or constructing meaning implies more than just changes in the nature of daily classroom activities. It also implies a different timetable for learning. Traditionally, most mathematics instruction was based on a mastery model in which we presented new information, provided children the opportunity to practice it, and generally expected most children to demonstrate mastery by the end of a unit—usually about three to four weeks of instruction. We then progressed to a new topic.

In contrast, the "learning as making meaning" viewpoint describes learning as a more gradual, growing proficiency in which students make sense of mathematical ideas as they use them to understand a variety of situations. Although a unit of instruction can provide focused experiences around particular concepts or skills, continued, regular use of this mathematics is essential for students to truly understand it and be able to use it flexibly in new situations.

In addition, there is growing evidence that informal use and exploration of concepts before more-focused instruction enhances learning (Bell 1990; Arons 1984). For example, regular informal experiences with equal sharing in grades K–2 appears to contribute to students' understanding of division in grade 3. Similarly, creating rules (formulas) to describe relationships in elementary and middle school are important informal experiences with variables that can enhance more-formal work in algebra.

Although students' abilities to solve many types of problems via informal means prior to formal instruction are well documented (e.g., Carpenter and Moser [1983]), the importance of informal experiences for future learning is just emerging.

Inherent in this changing understanding of understanding is the need for us as teachers to examine our philosophy of what mathematics is, what it means to know mathematics, and how we teach mathematics.

Beliefs about Mathematics

Another aspect of this view of learning is that what students learn is fundamentally connected to how they learn it. What students learn about particular procedures, as well as what view of mathematics and of themselves as learners of mathematics they develop, depends on the ways in which they engage in mathematical activity in their classrooms. Thom (1972) suggests that mathematical pedagogy reflects one's philosophy of mathematics. A similar view comes from Hersh's (1986, p.13) statement "One's conception of what mathematics is affects one's conception of how it should be presented."

This sends a powerful message to us as teachers. What philosophy of mathematics do students see in our classes? Is it coherent? Do we consciously try to make explicit matters having to do with what mathematics is? Do we engage students in activities that cause them to consciously reflect on their deep-seated beliefs about mathematics and what it means to know and to do mathematics? What implicit messages about mathematics are we sending though our actions and words?

In recent years research on teachers' beliefs and the interaction between beliefs and practice has received increasing attention. Thompson (1984) investigated high school teachers' beliefs and their classroom teaching and found evidence that teachers' beliefs, views, and preferences about mathematics influence what they do in the classroom. (Others who have studied teacher beliefs and the impact on teaching and learning are Shaw [1989]; Cooney [1985]; Brown [1985]; Dougherty [1990]; Peterson et al. [1989]; Schram et al. [1989]; Nespor [1987]; and Ernest [1988].)

We know from research that the deeply held beliefs of teachers about what can and should happen in school, about what is possible and what is desirable, and about the nature of understanding (Stigler and Perry 1988) are particularly difficult barriers to change. But we cannot change teaching unless we confront what we as teachers bring to teaching. We need to engage in professional conversations that will push our thinking, that will challenge our expectations for students, and that in the process will challenge what we believe and value about mathematics.

We need to help each other recognize the implicit messages that we send about the nature of mathematics and who can do mathematics. How are mathematical disagreements in the classroom resolved? On the basis of mathematical reasoning? Or is the teacher the voice of authority? Who is called on? Are the responses and questions of all students valued? Peer observation and feedback can provide valuable information about what we are communicating to our students unconsciously through our words and actions.

The kind of standards-based teaching that teachers are trying to accomplish all across the country amounts to nothing less than a complete paradigm shift in the classroom. No longer is the teacher the center of attention, the source of all knowledge. The teacher is now a poser of interesting problems, a guide, a questioner, whose role is to stimulate and steer the student on a quest to make sense of mathematics.

TEACHER DECISION MAKING AND REFLECTION

Teaching is a complex practice that is not reducible to recipes or prescriptions (NCTM 1991, p. 22). In this section, we take a close look at the important decisions that teachers make and the factors that must be considered in standards-based teaching.

Standards-based teaching typically is described as involving classrooms that are student-centered environments, in which students have the opportunity to engage in individual and group activities; use manipulatives and technology; and discuss, write about, and reflect upon their experiences. But is that an accurate characterization? Consider the following fourth-grade classroom scene:

> Students are working with partners, their desks pushed together and covered with color tiles, graph paper, markers, scissors, and paper. Some students are huddled together, making shapes with the tiles. Others are recording their color-tile shapes on graph paper, while others are sharing their findings with another partnership. Still others are writing in their journals. The teacher is circulating through the room, monitoring the progress of each pair. "Be sure to share your results with the pair across from you. We will have a group discussion in about five minutes."

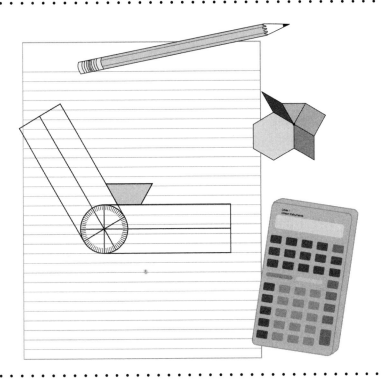

Is this how mathematics should be taught? It is tempting to answer a resounding yes because all the commonly cited components appear to be there—hands-on activities with manipulatives, group work, writing, discussion. But critical information is missing! What is the task in which the students are engaged? What mathematics is the task intended to elicit? And what mathematical thinking is it actually eliciting from students? What conceptions of mathematics are students forming as they do this task? First and foremost, the quality of mathematics instruction depends on the tasks in which in the students are engaged.

The reform of mathematics education is about *mathematics* and teaching and learning mathematics. Manipulatives, cooperative learning, discourse, writing, and authentic assessment are tools and strategies—powerful tools and strategies—for enhancing student learning. However, the presence of technology or cubes or pattern blocks does not guarantee that students learn. Putting students in groups and having more talk in the classroom does not necessarily mean that student learning is enhanced. An important mathematical question must be the center of the discussion and exploration. There are examples of very powerful learning from students in environments where new tools and techniques are being used.

But there also are examples of situations where the elements are there but their use is not centered on important mathematics, and little mathematics is learned. The quality of mathematics instruction depends on the mathematical thinking in which students are engaged and their resultant learning of mathematics. Choosing worthwhile mathematical tasks around which to organize instruction is the key to fostering quality mathematics instruction.

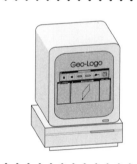

Worthwhile Mathematical Tasks

There is no other decision that teachers make that has a greater impact on students' opportunity to learn and on their perceptions about what mathematics is than the selection or creation of the tasks with which the teacher engages students in studying mathematics. The *Professional Standards for Teaching Mathematics* (NCTM 1991, p. 25) provides a number of criteria to consider in task selection and creation.

Standard 1: Worthwhile Mathematical Tasks

The teacher of mathematics should pose tasks that are based on—

* sound and significant mathematics;

- knowledge of students' understandings, interests, and experiences;
- knowledge of the range of ways that diverse students learn mathematics; and that
- engage students' intellect;
- develop students' mathematical understandings and skills;
- stimulate students to make connections and develop a coherent framework for mathematical ideas;
- call for problem formulation, problem solving, and mathematical reasoning;
- promote communication about mathematics;
- represent mathematics as an ongoing human activity;
- display sensitivity to, and draw on, students' diverse background experiences and dispositions;
- promote the development of all students' dispositions to do mathematics.

Here is a situation to consider (Connected Mathematics Project 1994a).

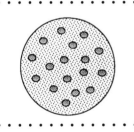

In the thousands of pizza restaurants across the United States, pizzas are sold in small, medium, and large sizes—usually measured by the diameter of the circular pie. Of course, the prices are different for the three sizes. Do you think a large pizza is usually the best buy?

The Sole D'Italia Pizzeria sells small, medium, and large pies. The small pie is nine inches in diameter, the medium pie is twelve inches in diameter, and the large pie is fifteen inches in diameter. For a plain cheese small pizza, Sole D'Italia charges $6; for a medium pizza, it charges $9; for a large pizza, it charges $12. Are these fair prices?

- Which measures should be most closely related to the prices charged—circumference or area or radius or diameter? Why?
- Use your results to write a report on the fairness of Sole D'Italia's pizza prices.

Let's look carefully at this task. Is it likely to be of interest to middle school students? Is the situation likely to be one that middle school students have encountered? The situation seems to pass both of these tests. Students eat pizza in all parts of the country. They know what the situation is about. They have likely wondered about the answer to the problem posed in the situation.

What about the mathematics that students would encounter in pursuing a solution to the problem? Clearly the problem would engage students in measuring circles in meaningful ways—radius, diameter, circumference, and area. Even the problem of producing a circle to represent each of the pizzas is an interesting challenge for many middle school students. A teacher might choose to use a problem of this type to lay the groundwork for needing efficient ways to find the area and the circumference of a circle.

Figures 1 and 2 are diagrams of student work on the task and a summary of the ways students in the class found to estimate the area of the circles representing the pizzas.

Figure 1

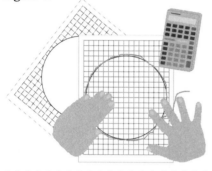

The students have certainly bumped into important mathematics in finding the data they need to make sense of the problem. Now the students have an opportunity to use these data to reach a conclusion about the situation. They will write about their work and what they think is a fair way to price pizzas. They will take from the situation and the problem posed useful mathematical experiences and techniques, as well as useful ideas to use in consumer situations that they encounter in the world.

From the perspective of the teacher, this situation has laid the groundwork for the need for more systematic methods of measuring circles. Embedding this problem in a series of problems that expose students to more and more information or data about circles can lead to the realization that it takes a little more than three squares with the radius as an edge to cover a circle regardless of the size of the circle. You can see that the students in their work on the pizza problem are seeing the symmetry on the circle that they have drawn on a square grid. They are already looking at the circle and the surrounding square in quarters. The move from here to using "a little more than three radius squares" as an

Figure 2. Area

1. Count all the squares and parts of squares.

2. Find 1/4 of the area. Multiply by 4 to get your answer. (Jimmy) (Jill)

3. Put a square around the circle. Find the area of a small outside part. Multiply by 4 to get all the outside area. Find area of the square. Subtract the outside area to get the area of the circle. (Jimmy)

estimate of the area of a circle, and then to the formula $A = \pi\, r^2$ is a matter of more experience with circular situations and skillful questioning on the part of the teacher.

Embedding worthwhile tasks in a series of problems that support students' construction of mathematical ideas is a critical aspect of effective instruction. Tasks must be sequenced so that students "bump into" important mathematics, make their own conjectures and theories, then modify and refine their understandings as they further investigate these ideas and attempt to address new situations.

For example, although the pizza problem engages students in using important mathematical ideas, its instructional value comes from the meaningful basis it provides for subsequent systematic investigations of the area and perimeter of circles. Moving from the pizza problem to another, unrelated problem, no matter how worthwhile, would not provide students with the opportunity to capitalize upon what they had discovered through the pizza investigation.

Designing an effective sequence of tasks requires an analysis of both the mathematics that the students are to understand and students' current understanding of that mathematics. An analysis of the mathematical concepts and skills to be understood and their prerequisites and connections to other mathematical ideas is essential for determining an overall task sequence that will engage students in thinking about the important mathematical ideas. Analysis of students' emerging understandings, difficulties, and misconceptions enables the teacher to select and sequence tasks that will help students develop more sophisticated understandings. (We will return to this point in the Classroom Discourse and Analysis and Assessment sections.)

Mathematics learning is further enhanced when tasks provide students opportunities to use previously learned concepts and techniques in the process of discovering new ones. For example, several days after doing the pizza problem, the teacher had her students do the standard investigation of the relationship between the circumference and the diameter of circles. Each pair of students measured the circumference and diameter of a small number of objects, then computed various arithmetic relationships between them, for example, sum, difference, product, ratio.

Instead of stopping there, the teacher recognized this as an opportunity for students to use data analysis techniques to estimate the value of pi. The class made a line plot of all the values obtained for circumference/diameter (see fig. 3), then students discussed what they saw. The discussion centered on sorting out the extreme values.

Sometimes extreme values tell the best part of the story. Here the teacher asked questions about which of the data points seemed reasonable and which seemed problematic. The students decided that the measures that produced the values in the intervals 2.2, 2.4, 3.6, 4.0, and 4.4 should be reexamined. Measurement and computation errors were found and the data were cleaned. The teacher asked the class members what they would report if they could give only one number as their estimate for the relationship between the circumference and the diameter.

· ·

Figure 3

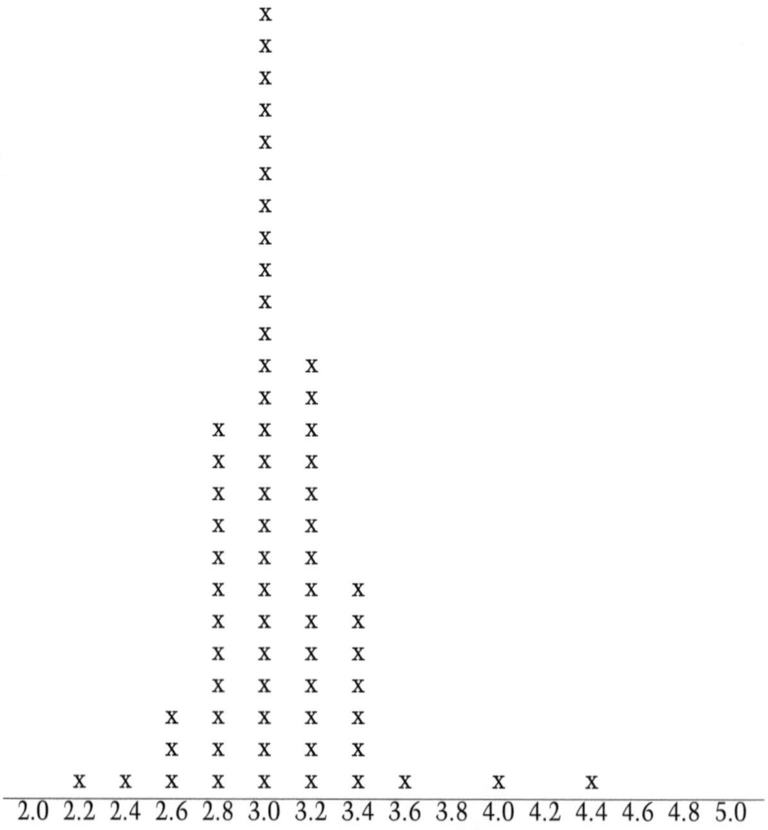

```
                              X
                              X
                              X
                              X
                              X
                              X
                              X
                              X
                              X
                              X
                              X
                              X   X
                              X   X
                      X       X   X
                      X       X   X
                      X       X   X
                      X       X   X
                      X       X   X
                      X       X   X   X
                      X       X   X   X
                      X       X   X   X
                      X       X   X   X
              X       X       X   X   X
              X       X       X   X   X
      X   X   X       X   X   X   X   X       X           X
  ─────────────────────────────────────────────────────────────
   2.0 2.2 2.4 2.6 2.8 3.0 3.2 3.4 3.6 3.8 4.0 4.2 4.4 4.6 4.8 5.0
```

· ·

The students noticed that the values tended to clump between 3 and 3.4. The class decided that the mean of the values between 3 and 3.4 was the best estimate of the ratio. Using this technique, students took the actual data (not the grouped data for the line plot) and estimated pi as 3.143. By encouraging the students to use the data analysis techniques they had learned earlier in the year in this investigation, the teacher gave them an opportunity to review these techniques and see their usefulness.

Worthwhile mathematical tasks have many different forms and durations. They can range from short questions posed by the teacher or student to investigations like the pizza problem described above that take one or more class periods to projects requiring several weeks of work. They may involve students working in small groups, in pairs, or independently.

Games are often overlooked as worthwhile mathematical tasks, yet they can be powerful tools for introducing new concepts and practicing skills (Bright et al. 1985; Kamii and Lewis 1990; and Randel et al. 1992). For example, Circles and Stars is a two-person game that gives children a visual representation of multiplication as repeated addition (Burns 1991). Children first roll a die to determine how many circles to draw, then roll again to find out how many stars to draw in each circle. The winner is the child with the most stars after seven rounds.

Using standard multiplication notation (e.g., $3 \times 4 = 12$) to describe each round introduces the notation in a meaningful context and illustrates that the numbers in the equation represent different quantities; that is, the first number represents the number of circles (groups), the second represents the number of stars in each group, and the third represents the total number of stars. At the high school level, computer games like Green Globs (Dugdale and Kibbey 1985) can create a need for students to know more mathematics, for example, how to create equations for sloped lines or for curves.

Games are also effective contexts for drill and practice. From simple dice and card games at elementary levels to calculator games to computer games at the secondary level, games engage students in repeatedly using mathematics to accomplish tasks that are meaningful to them, thus providing the repeated exposure necessary for basic facts, concepts, and skills to become automatic.

The criteria for worthwhile instructional tasks also apply to creating and selecting practice and homework activities. Research indicates that practice is critical for skills and knowledge to become automatic. Clearly, there are a number of skills that students need to have automatically at their disposal if they are to be powerful users of mathematics. But practice can, and should, occur through worthwhile, engaging activities. Homework assignments like asking grade 1 students to find as many things as they can in their home that are six inches long (Bell 1990) or asking grade 4 students to play the Product Game (Lappan et al. 1983) with a family member or friend provide practice in fundamental skills yet are more interesting and engaging than traditional paper-and-pencil worksheet activities.

Along with the selection of worthwhile tasks, the *Professional Standards for Teaching Mathematics* (NCTM 1991) portrays a very different kind of classroom interaction about the mathematics and a very different set of tools, both physical and intellectual, for doing mathematics.

CLASSROOM DISCOURSE

The *Professional Teaching Standards* describes discourse as "the ways of representing, thinking, talking, agreeing, and disagreeing" (p.36) as a group of students and a teacher strive to make sense of mathematics. Mathematical discourse includes the accepted rules of evidence and the methods of argument that are used to validate mathematical ideas and procedures. Discourse includes the ways that ideas are represented, exchanged, and modified into more powerful and useful ideas.

Teachers have a critical role to play in establishing the norms of discourse in the classroom and orchestrating discourse on a daily basis. It is through the interactions in the classroom that students learn what mathematical activities are acceptable, which need to be explained or justified, and what explanations or justifications are acceptable.

Organizing Students for Instruction

The emphasis on discourse in the *Professional Teaching Standards* should not be interpreted as a stand only for group work in a mathematics classroom. Students need opportunity and encouragement to think and struggle on their own to make sense of mathematics. But they also need to have ample opportunities to work cooperatively with other students who bring different ideas to the task at hand.

The classroom environment needs to be one where a student feels comfortable in giving and in seeking advice, help, or challenge from another student or the teacher. In addition to individual and group activities, the teacher and the class need opportunities to synthesize ideas and strategies that the group of students working together, or individually, have used or invented.

Here is an example that focuses on this synthesis activity of the whole class. The teacher, Ms. M, and her students have been engaged in mathematical task situations that require the students to approximate the area of irregular shapes. The students have used covering and counting as a strategy for making such estimates of area. The shapes are put onto grid paper or covered with a transparent grid and the students then find ways to "count."

The teacher's primary mathematical goal is building a deep understanding of what it means to measure area. She wants the students to see that area is a "covering" and that the goal of any procedure for finding area is to make the counting of the square units needed to cover the area as efficient as possible. The teacher wants to know what sense her students have made of the problems they have investigated over the past several days. She assigned a task she had gotten from a mathematics solutions workshop of Marilyn Burns Education Associates as an embedded assessment. The task felt no different to the students from the tasks the teacher had used to develop the ideas about measuring irregular shapes.

The teacher gave each student the same hand, drawn on centimeter grid paper (fig.4), and asked each to find the area of the hand in square centimeters.

Figure 4

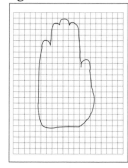

The teacher could have taken these home and graded them, written a number down in her grade book, and returned the papers. She might have little insight from this about what and how the students are thinking about these kinds of situations, what the students understand, and what naive conceptions they hold. Because knowing this was the focus of the assessment for this teacher, she took class time to have a whole-group discussion of the class results.

The teacher and the students made the stem-and-leaf plot in figure 5 to show what each student in the class had found.

Figure 5. Stem-and-Leaf Plot of Class Data

9	9
10	
11	
12	6 5 5 5 5
13	3 7 7 9 5 6 9 1 2 3 3 7 5 9 3 5 9 1 0
14	2 9 2 9 0
15	
16	
17	
18	
19	
20	0

From the picture of the class data, the approximate nature of measurement is seen clearly. Most of the estimates fall into a cluster of data points that range from 126 cm² to 149 cm². However, the two outliers, 99 cm² and 200 cm², stimulated a rich class discussion that helped the teacher understand where her students were. The classroom discourse also helped the students deepen their understanding of what it means to measure area and how area is or is not related to perimeter. Here is a bit of the classroom discourse:

Ms. M: Let's look at the two outliers. Who got these estimates?

Estella: The 99 cm² is mine and I added wrong! I had missed adding on some of the parts of the hand. I added all of the parts again and got 132 cm².

Ms. M: Let's make that change on the stem-and-leaf plot. Who got the 200 cm²?

Lee: I did.

Ms. M: Would you explain how you thought about the problem? What strategy did you use for approximating the area?

Lee: I, um, well I started trying to count all of the squares and I kept losing track. So I tried to think of an easier way. And then I used a string.

Ms. M: (To the class.) Is there anything you want to ask Lee?

Sadie: Yeah! I don't see how you used a string.

Lee: Well, I cut the string to go around the edge of the hand. Then I held the string together at the ends and sort of squared it up. It's easier to find the area that way. Squares are easy.

Ms. M: What do you think of Lee's method? Will it give a good estimate of the area?

Students: Cool!

Estella: It seems too big to me. Something is wrong!

From this discussion the teacher has learned that her students are still struggling to make sense of the relationship between area and perimeter. Most of the class members have reasonable skill at using covering and counting to find an irregular area. But several students have not sorted out the underlying relationship between area and perimeter. Lee's method, which does not give reliable results, seems compelling to several students.

Now the teacher has to decide what questions to ask or tasks to explore that will push the thinking of her students beyond the naive conception shown in Lee's method. She chose to pose another task to be explored at home. Each student made two drawings of his or her hand: one with the fingers together, as was the case in the assessment task, and another with the hand drawn on a grid with the fingers apart.

She chose this approach because she felt that her students were pretty solid on their understanding that area stays the same whether the print is fingers touching or fingers separated (with the exception of the bits of weblike skin that stretches between the fingers and may increase area slightly). Therefore, the data from Lee's method on this task would raise some dissonance that would have to be resolved.

In addition, the teacher chose another task for the following day that engaged the students in cutting and rearranging the area of a square to create figures with much larger perimeters. In this example, we see a teacher using classroom discourse to push students' mathematical thinking, as well as to provide data for making instructional decisions.

Problem Centered Teaching

Changing the nature of both the mathematical tasks posed and the discourse norms in the classroom creates new classroom roles for teachers. If students are to have opportunities to explore rich problems within which the mathematics will be confronted, the teacher has to learn how to be effective in at least four new roles: (1) engaging the students in the problem, (2) pushing student thinking while the exploration is proceeding, (3) helping the students to make the mathematics more explicit during whole-class and group interaction and synthesis, (4) using and responding to the diversity of the classroom to create an environment in which all students feel empowered to learn mathematics.

A reflective teacher realizes that engaging the students in a problem does not mean just having fun with the context. It is important, of course, for students to understand the context. But having a mathematical question in the students' heads when they explore the situation is critical.

The teacher has to work with the students to help them understand the question(s) that mathematics can help answer in the situation. This means keeping an eye on the goal in posing the task. This does not mean that the teacher structures the mathematical question(s) so that no thinking or work is left to the student. It means that the teacher keeps the focus on the big question embedded in the task and uses her judgment about whether in this instance the students should formulate the questions for themselves or should find answers to problems in the situation that are posed by the situation or the teacher.

As the students work on the task posed—often in groups, always using tools such as calculators, computers, and physical manipulatives as well as intellectual tools such as analogies—the teacher has the role of assessing what sense her students are making of the task and the mathematics. She circulates among the groups asking questions, asking for evidence to support students' conjectures, redirecting and refocusing groups that are off task or floundering. Here the teacher is a coach, a guide, an interlocutor, and an assessor of student progress and problems.

After most of the groups have made progress on the problem, the teacher and the class come together to look at the different answers or the different data collected, to look at the strategies used, and to examine the conjectures the groups have made and the evidence the groups have to support the conjectures.

It is in this group sense-making role that the teacher must keep his or her eye on the mathematical goals embedded in the task, on bringing the mathematics alive and making it more explicit and powerful for the students, on helping students to connect what they have learned to things they already know. This is where the teacher works to set higher expectations for both the work of students and the ways in which students are to engage in discussions with each other and the teacher and class together. Changing classroom norms is very hard. Students are resistant. They know what mathematics class should be like and even if they do not enjoy mathematics, they will clamor for more directions, less thinking on their own, and for task completion to be the goal of the class. However, the compelling evidence from teachers trying to make change toward the *Professional Teaching Standards* (NCTM 1991) is that if a teacher perseveres in setting higher expectations and insisting that students reason and defend their conclusions, students will respond.

The roles of reaching all students and of creating a classroom environment supportive of the standards are intertwined. Students must feel that they have a stake in this class. They must feel that their ideas are valued and that regardless of socioeconomic status, race, color, or language, the teacher believes in them, believes that they can learn, and expects no less of one student than of any other student in the class. If we are to reach all students, this must be our commitment: to create classroom environments in which the ethos of the classroom supports students to feel that it is safe to have ideas, to express them, to ask questions, to question others, to ask for help, and to give help. Yet in such an environment of standards-based teaching, the teacher faces the dilemma of balancing telling students (and when and how to tell them) with leaving them to struggle on their own.

The Role of Teacher "Telling"

Expecting students to discover centuries of mathematical conventions and powerful techniques is not what classroom discourse is about. Making sense of mathematics surely will involve the teacher at times telling students things about mathematics. Reform is about posing interesting and challenging problems that will create a situation where the use of mathematics is empowering. Reform is about letting students have the time to explore these problems individually and with their classmates.

We need to understand that what students have worked hard to discover on their own belongs to them. It can be recreated and used when the need arises. And yet, teachers must make reasonable decisions about what is "discoverable" and what needs to be shared from the commonly accepted body of mathematical knowledge and convention. This dilemma of what and when to "tell" students about mathematics or about the correctness of their responses seems to be tied up in the expectations we set for students.

Expectations for Students

If we believe that students can think and reason given the proper opportunities and if we believe that our students are best served by a challenging mathematics curriculum requiring effort and persistence, then we have to organize our instruction around tasks that require thinking and sense making. And we have to be willing to support our students with our faith in them and our encouragement as they struggle not only with the mathematics but also with new expectations and requirements for how students and the teacher are to work within the classroom. This means that we will experience frustration at the time it takes for students to discover or reason their way through the mathematics embedded in tasks posed. We will want to hurry the process along by telling the students what they should be seeing rather than asking questions that stimulate, encourage, and guide students toward making their own sense of things.

Teachers who are working to set new expectations and norms for mathematics classrooms have very interesting stories to tell of their personal struggles to handle the dilemma of when to tell. In a sixth-grade classroom in Michigan, the students were challenged to work in their groups on the following problem from a Connected Mathematics Project (1994b) unit on data analysis:

> You know that a class of twenty-five students in a nearby school has a mean number of three children in their families. Work with your group to make a display showing one possible distribution of the numbers of children in each of the twenty-five students' families so that the mean number of children is three. Use stick-on notes to represent the families of each of the twenty-five students.
>
> a. When you have agreed on a distribution, explain how you know what you did works.
>
> b. Do you think that your group's display will be identical to or different from the displays of other groups in your class? Explain your reasoning.
>
> c. Is it possible to have some stick-on notes at the mean? No stick-on notes at the mean? Why or why not?

Most of the groups of students started their models by placing all twenty-five of the stick-on notes on the mean of three children per family and moving pairs of notes away from the mean in opposite directions. The groups of students presented their results to the class, and during the process of presenting, the teacher and the class asked questions of the group at the board.

The question of whether or not there had to be a stick-on note on the mean was surprisingly hard for the students. In every group the students argued that whether there had to be data at the mean depended on whether there were an odd or an even number of data points. With twenty-five students' families most of the students said that there had to be data at the mean—and all of the examples presented had data at the mean!

At this point the teacher could have intervened. She could have produced an example of a distribution with the number of children in twenty-five families and a mean of three children per family that had no family at the mean. This would have told the students that they were wrong. She instead asked the students to test their ideas at home that night by taking a smaller number of students—for example, nine students for one display and ten students for another—to see if their conjectures held up. She encouraged them to try to find several different distributions of numbers of children in families for the nine- or ten-student class, but in every case to make the mean three children in a family.

The teacher was interviewed after the class and shared her concern that she had spent so much time on a point that she did not think would be so hard for the students. She had assumed that the distributions produced by the groups of students would include an example to cause them to think more deeply about what a mean is. However, the teacher was committed to helping her students develop a deep understanding of major concepts and she considered the first engagement with mean as a measure of center to be very important. She wanted the students to understand that the mean is a balance point for the distribution, not just an algorithm that produces a number.

The next day the class returned to the questions they had struggled with the day before. This time several students had managed to create distributions that had families at the mean and other distributions with no family at the mean for both the nine- and ten-student situations! Figure 6 shows some examples.

One of the students in the class, David, said that he found that you had to have twenty-seven children total in all of the families for the nine-student example and thirty children total for all of the families of the ten-student example. There was quick agreement that this made sense. Dianna made a real breakthrough when she explained that she found that whenever she moved a family (a stick-on note) to the right, she had to move a stick-on note the same number of spaces to the left. She said that she initially thought that if you moved a note two spaces to the left of the mean, you had to move a note two spaces to the right to balance.

Figure 6

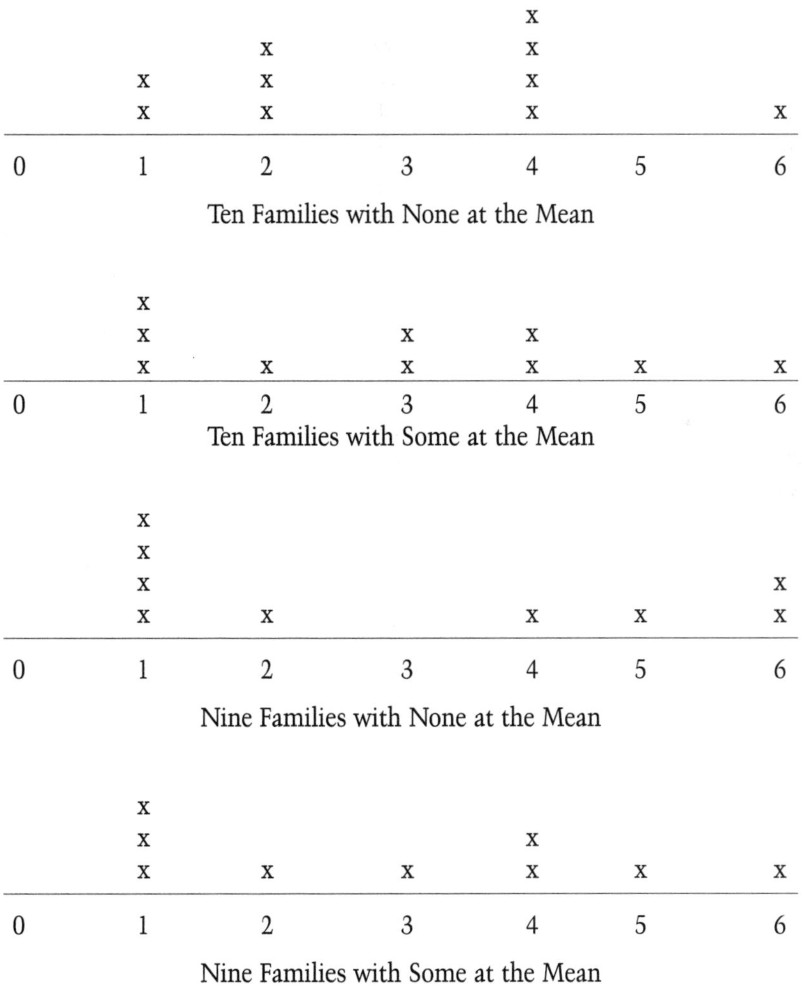

Ten Families with None at the Mean

0	1	2	3	4	5	6
				X		
		X		X		
	X	X		X		
	X	X		X		X

Ten Families with Some at the Mean

0	1	2	3	4	5	6
	X					
	X		X	X		
	X	X	X	X	X	X

Nine Families with None at the Mean

0	1	2	3	4	5	6
	X					
	X					
	X					X
	X	X		X	X	X

Nine Families with Some at the Mean

0	1	2	3	4	5	6
	X					
	X			X		
	X	X	X	X	X	X

Then she figured out that you could move *two different notes one space to the right* of the mean and still have balance. Out of this discussion nearly all of the class saw that David's observation made sense and many of the students understood what Dianna was saying and began checking out other examples.

By no means was the teacher's dilemma solved. The discussion had taken time. And some of the students still were not convinced by Dianna's observation. David's observation helped most of the class toward a more efficient way to find the mean. But this method does not support understanding the balance interpretation of the mean as a point in the distribution where the total *dis-*

tances from the mean of the stick-on notes on each side is the same. Dianna's observation pushed toward this more powerful understanding about the mean.

Would it have been more efficient for the teacher to tell the students an algorithm for finding the mean and then have them practice? Of course, if our goal is to have a way to find a thing called the mean. However, if our goal is mathematical thinking and deep understanding, the students must have time to make sense of situations through active mental and, at times, physical involvement.

ANALYSIS AND ASSESSMENT

Throughout our examples of standards-based teaching, teachers have been reflective decision makers, monitoring their own practice and the learning of their students. They observe and listen to students, gathering evidence about what students are thinking and the sense that they are making of mathematics. Teachers have used this information to analyze their instruction and make decisions about what to do next, either immediately in their orchestration of classroom discourse or in subsequent lessons (e.g., area of a hand investigation). Their assessment is embedded in instruction. Its goal is to support the learning of each student by gathering evidence about each student's understanding. They use this information to guide their own instructional decisions and/or to monitor the progress of students. Such practices illustrate the vision of assessment that is part of standards-based teaching described in the *Assessment Standards* (NCTM forthcoming) and the *Professional Teaching Standards* (1991).

The *Assessment Standards* describes assessment as a process of gathering evidence about a student's mathematical power and of making inferences based on that evidence for a variety of educational purposes. Assessment is thus one of the greatest challenges in teaching. In essence, we are in the position of the man in the cartoon in figure 7. We are trying to ascertain each student's understanding based on what we can actually observe.

Assessment involves a complex set of decisions. The first decision is what to assess. The mathematics to be assessed should represent the teacher's mathematical goals for students. The next decisions are what evidence to gather (e.g., observations, students' responses to questions, and the products of student work) and what assessment methods to use (e.g., performance tasks, projects, and group or individual work). Finally, decisions must be made about how to use the evidence to make judgments about students' learning and about instructional practices.

The *Assessment Standards* identifies key criteria for making these decisions. First, since what we assess and how we assess it communicate what we value, in standards-based teaching assessment reflects the mathematics that

Figure 7

Harold Asturias

is most important for students to learn. Assessment provides evidence about the mathematics that is valued—problem solving, reasoning, conceptual understanding—rather than what is easy to assess (e.g., execution of procedures).

Second, assessment enhances learning by being part of ongoing classroom activity, not an interruption to the instructional process. And, to enhance the learning of each student, assessment is equitable, that is, each student has the opportunity to demonstrate his or her understanding. This may mean providing different students with different ways to show what they know and can do or considering students' perspectives in judging responses.

Students also understand the expectations and process of assessment in the classroom. They know what they are expected to know and the criteria for judging their work. In fact, students as well as teachers are assessors in the classroom. They have the opportunity to review and make judgments about the quality of their work and their progress as mathematics learners.

For assessment to support the learning of each student, the evidence gathered must be used to construct valid inferences about each student's understanding. The key question for teachers is, "What does this information—taken in conjunction with other information about this student—tell me about the student's understanding of mathematics?" Too often, the major concern is *scoring* student responses, that is, developing scoring guides or rubrics, putting numbers on students' work. Though such concerns often reflect classroom realities, the essential instructional issue is the inferences that can be made about the nature of students' understandings and their progress in making sense of mathematics.

SUMMARY

Teachers at all levels are finding ways to get started toward the vision of teaching and learning portrayed in the *Professional Standards for Teaching Mathematics* (NCTM 1991). More important, they are finding the will to continue long enough to see a changed mathematics program become the norm in the school.

The obstacles are formidable. We can stay at the stage of reciting why we cannot change, or we can begin to slowly and deliberately wield whatever influence and skill we have to give change a push. New curriculum materials are being developed at all levels, new assessment instruments are being designed, schools are experimenting with new structures, the National Board for Professional Teaching Standards is moving toward a certification process to help raise teaching to a professional status.

Other disciplines are following the lead of mathematics to articulate standards for curriculum, teaching, and evaluation. The question is, Will we be the ones that push these efforts by building the climate and support for change over time, or will we be the impediments who are locked in our own paradigms for teaching and learning, afraid to try new ideas?

Some of our efforts will fail. But we can learn from these as we continue to move forward. We are all bound by the same moral imperative: to do the best we can for the children in our communities. We must recognize and take comfort in the knowledge that these are truly promising times in mathematics education. The doors of our mathematics classrooms have been opened both in K–12 and in higher education, and it is unlikely that we could get them closed again—even if we wanted to.

• • • • • • • • • • • • • • •

BIBLIOGRAPHY

Arons, A. B. "Education through Science." *Journal of College Science Teaching* 13 (4) (1984): 210, 212–20.

Bell, Max. *Everyday Mathematics, Grades 1, 2 and 3*. Evanston, Ill.: Everyday MathTools Publishers, 1990.

Bright, George W., John G. Harvey, and Margariete Montague Wheeler. *Learning and Mathematical Games. Journal for Research in Mathematics Education* Monograph 1. Reston, Va.: National Council of Teachers of Mathematics, 1985.

Brooks, J. G., and M. G. Brooks. *The Case for Constructivist Classrooms*. Alexandria, Va.: Association for Supervision and Curriculum Development, 1993.

Brown, Catherine. "A Study of the Socialization to Teaching of a Beginning Mathematics Teacher." Ph.D. dissertation, University of Georgia, Athens, 1985.

Brown, John Seeley, Allan Collins, and Paul Duguid. "Situated Cognition and the Culture of Learning." *Educational Researcher* 18 (1) (1989): 32–42.

Bruner, Jerome. "Science Education and Teachers: A Karplus Lecture." *Journal of Science Education and Technology* 1 (1) (1992): 5–12.

Burns, Marilyn. *Math by All Means: Multiplication, Grade 3.* White Plains, N.Y.: Marilyn Burns Education Associates, 1991. Distributed by Cuisenaire Co. of America.

Carpenter, Thomas P., Elizabeth H. Fennema, P. L. Peterson, C. P. Chiang, and Megan Loef. "Using Knowledge of Children's Mathematics Thinking in Classroom Teaching: An Experimental Study." *American Education Research Journal* 26 (4) (1989): 499–531.

Carpenter, Thomas P., and James M. Moser. "The Acquisition of Addition and Subtraction Concepts." In *Acquisition of Mathematics: Concepts and Processes*, edited by Richard Lesh and Marsha Landau. New York: Academic Press, 1983.

Carraher, Terezinha N., David W. Carraher, and Analúcia D. Schliemann. "Mathematics in the Streets and in the Schools." *British Journal of Developmental Psychology* 3 (1) (1985): 21–9.

Coldiron, J. R. "Desired Outcomes, Alternative Assessment, and Chapter 1: An Overview of the Issues." Paper presented at the Pennsylvania Desired Outcomes and Alternative Assessment Conference, King of Prussia, Pa., February 1994.

Connected Mathematics Project. "About Us." Draft. East Lansing, Mich.: Michigan State University Connected Mathematics Project, 1994b.

_____. "Covering and Surrounding." Draft. East Lansing, Mich.: Michigan State University Connected Mathematics Project, 1994a.

Cooney, Thomas J. "A Beginning Teacher's View of Problem Solving." *Journal for Research in Mathematics Education* 16 (November 1985): 324–36.

Dougherty, B. J. "Influences of Teacher Cognitive/Conceptual Levels on Problem-Solving Instruction." In *Proceedings of the Fourteenth International Conference for the Psychology of Mathematics Education*, edited by George Booker et al., pp. 119–26. Oaxtepec, Mexico: International Group for the Psychology of Mathematics Education, 1990.

Dugdale, Sharon, and David Kibbey. Green Globs and Graphing Equations. Pleasantville, N.Y.: Sunburst Communications, 1985. Software.

Ernest, Paul. "The Impact of Beliefs on the Teaching of Mathematics." Paper prepared for International Congress on Mathematics Education VI, Budapest, Hungary, July 1988.

Hersh, Ruben. "Some Proposals for Revising the Philosophy of Mathematics." In *New Directions in the Philosophy of Mathematics*, edited by Thomas Tymoczko, pp. 9–28. Boston: Birkhauser, 1986.

Kamii, Constance, and Barbara Ann Lewis. "Research into Practice: Contructivism and First-Grade Arithmetic." *Arithmetic Teacher* 38 (1) (September 1990): 36–7.

Knoblauch, C. H., and Lil Brannon. *Rhetorical Traditions and the Teaching of Writing.* Upper Montclair, N.J.: Boynton Cook Publishers, 1984.

Lappan, Glenda T., W. M. Fitzgerald, E. D. Phillips, and M. J. Winter. *Factors and Multiples.* Menlo Park, Calif.: Addison Wesley Publishing Co., 1985.

Morrow, Charlotte. "A Description of SummerMath, a Process Oriented Mathematics Learning Community and Resulting Student Attitudes toward Mathematics." Paper presented at the National Council of Teachers of Mathematics Annual Meeting and Research Presession, New Orleans, La., April 1991.

National Council of Teachers of Mathematics. *Assessment Standards for School Mathematics.* Reston, Va.: The Council, forthcoming.

———. *Curriculum and Evaluation Standards for School Mathematics.* Reston, Va.: The Council, 1989.

———. *Professional Standards for Teaching Mathematics.* Reston, Va.: The Council, 1991.

Nespor, Jan. "The Role of Beliefs in the Practice of Teaching." *Journal of Curriculum Studies* 19, pp. 317–28.

Randel, J. M., B. A. Morris, C. D. Wetzel, and B. V. Whitehill. "The Effectiveness of Games for Educational Purposes: A Review of Recent Research." *Simulation and Games* 23 (3) (1992): 261–76.

Schram, Pamela, Sandra Wilcox, Glenda Lappan, and Perry Lanier. "Changing Preservice Teachers' Beliefs about Mathematics Education." In *Proceedings of the Eleventh Annual Meeting of the International Group for the Psychology of Mathematics Education, North American Chapter,* edited by Carolyn Maher, pp. 349–55. New Brunswick, N.J.: Rutgers University, 1989.

Shaw, K. "Contrasts of Teacher Ideal and Actual Beliefs about Mathematics Understanding: Three Case Studies." Unpublished doctoral dissertation, University of Georgia, 1989.

Stigler, James, and Michelle Perry. "Cross-Cultural Studies of Mathematics Teaching and Learning: Recent Findings and New Directions." In *Perspectives on Research on Effective Mathematics Teaching,* Vol. 1, edited by Douglas Grouws and Thomas Cooney, pp. 194–223. Hillsdale, N.J.: Lawrence Erlbaum Associates, 1988.

Thom, Rene. "Modern Mathematics: Does It Really Exist?" In *Developments in Mathematical Education,* edited by A. G. Howston, pp. 194–209. Cambridge: Cambridge University Press, 1972.

Thompson, A. G. "The Relationship of Teachers' Conceptions of Mathematics and Mathematics Teaching to Instructional Practice." *Educational Studies in Mathematics* 15 (2) (1984): 105–27.

Teacher Preparation

Jane O. Swafford

T HE preparation of teachers of mathematics is a critical link in reforming the way mathematics is taught in schools. The impact of the current reform movement on the preparation of teachers can best be understood within the historical context of the evolution of teacher preparation over the past seventy-five years. As critical issues emerge, promising innovative programs are being developed to address them.

HISTORICAL OVERVIEW: THE PAST SEVENTY-FIVE YEARS OF MATHEMATICS TEACHER PREPARATION

When the National Council of Teachers of Mathematics (NCTM) was founded in 1920, programs for the preparation of teachers in the U.S. were already nearly 100 years old. The first private academy for the preparation of elementary teachers was founded in Vermont in 1825, and the first state normal school was opened in Massachusetts in 1839 (Gibb, Karnes, and Wren 1970). By the turn of the century, the normal school movement had spread throughout the country and was the principal means for the preparation of elementary school teachers.

Initially, teachers for the secondary school were college and university graduates with no special preparation for teaching. As the normal school movement took hold as the agency for the preparation of elementary teachers, colleges and universities began to form departments and schools of education with the mission to prepare high school teachers.

By 1920, the typical two-year program of the normal school for the preparation of elementary teachers consisted of academic subject matter, profes-

· · · · · · · · · · · · · ·

Jane Swafford is a professor of mathematics at Illinois State University, where she works with preservice elementary and middle school teachers, in-service teachers, and future teacher educators. She is especially interested in the development of algebraic reasoning and in ways to keep more students in the algebra pipeline.

sional education, and supervised teaching in a laboratory school. The mathematics portion of the academic subject matter in the training of these early teachers for elementary schools focused on a review of arithmetic with some algebra and geometry. The goal of the program was to see that prospective elementary teachers knew the rules of computation and how to conduct computational drill.

The preparation of mathematics teachers for high schools was a four-year university program, consisting of a liberal arts major or possibly a minor in mathematics and some courses in the teaching of mathematics taken from the new departments or schools of education.

During the twenty years from 1920 to 1940, the Mathematical Association of America (MAA) took the lead in making recommendations about the preparation of mathematics teachers, with reports published in 1923 and 1935 (MAA 1923, 1935). In 1940, NCTM joined the chorus recommending improvements in teacher education with the publication, jointly with the MAA, of the Fifteenth Yearbook (NCTM 1940) on the place of mathematics in the secondary schools. This was followed by NCTM's participation in the Commission on Post-War Plans, whose reports were published in 1944 and 1945 (NCTM 1944, 1945).

By the late 1950s the typical elementary teacher had no actual college mathematics course. His or her preparation consisted of perhaps one year of high school mathematics and a course in the methods of teaching arithmetic. The typical secondary mathematics teacher had a minimal mathematics major consisting of precalculus, calculus, theory of equations, college geometry, differential equations, and two to three electives (Dubish 1970). Teachers of junior high generally were not recognized as a special category.

With the appearance of the "new math" curriculum materials in the late 1950s, the content preparation of teachers came under attack. In response, the MAA's Committee on the Undergraduate Program in Mathematics (CUPM) published *Recommendations for the Training of Teachers* (CUPM 1961, revised 1966). For prospective elementary teachers, CUPM recommended at least two years of college preparatory mathematics in high school and a twelve-semester hour mathematics sequence in college including number systems, algebra, and geometry. For the prospective junior high teacher, the recommendation was three years of college preparatory mathematics and seven courses or twenty-one semester hours in college mathematics including elementary analysis, abstract algebra, geometry, and probability. For the prospective high school teachers, a major in mathematics with a minor in a field using mathematics was recommended.

The impact of the CUPM recommendations was for institutions to move from requiring no mathematics of their elementary majors to requiring up to two courses. Few institutions, however, adopted the full twelve-hour sequence, and the content of the required courses lacked the depth suggested by CUPM (Dossey 1981). At the secondary level, the CUPM mathematics content recommendations were widely accepted and implemented with the exception of the second course in geometry. Few institutions introduced special programs for junior high teachers.

The 1970s saw little change in programs for the preparation of teachers of mathematics. The back-to-basics movement had sent innovation into hibernation. Outside of mathematics education, a growing national concern for the inadequacies of U.S. schooling was sparked by *A Nation at Risk* (National Commission on Excellence in Education 1983). With it, the first wave of educational reform was underway.

IMPACT OF THE REFORM MOVEMENT: NEW PREPARATION FOR A NEW VISION OF MATHEMATICS TEACHING

The reform movement erupted on the mathematics education scene with NCTM's release in 1989 of the *Curriculum and Evaluation Standards for School Mathematics*. The *Curriculum and Evaluation Standards* presented an entirely new vision of school mathematics. It calls for the development of mathematical power for all students, not just the elite. Rather than simple proficiency with computation and symbol manipulations and the memorization of facts, all students should develop the ability

- to explore, conjecture, and reason logically,
- to solve nonroutine problems,
- to communicate about and through mathematics, and
- to connect ideas within mathematics and between mathematics and other intellectual activities.

Further, students should develop a disposition to seek, evaluate, and use quantitative and spatial information in solving problems and making decisions.

Implicit in the *Curriculum and Evaluation Standards* is a new view of the teaching of mathematics that calls for teachers who are proficient in

- selecting mathematical tasks to engage student interest and intellect,
- providing opportunity to deepen understanding of mathematics and its applications,
- orchestrating classroom discourse in ways to promote investigation and growth of mathematical ideas,
- using and helping students use technology,
- seeking connections to previous and developing knowledge,
- guiding individual, small-group, and whole-class work, and
- continually assessing students' learning using a variety of methods.

This new vision of school mathematics and its teaching calls for a new vision for the preparation of teachers of mathematics and raises a number of critical issues.

CRITICAL ISSUES IN TEACHER PREPARATION

How can we best prepare prospective teachers to teach in a manner consistent with the reform movement? What sort of knowledge, beliefs, and practices must prospective teachers develop so that as new teachers they can enter the classroom able to implement and maintain the new vision of school mathematics?

As a companion to the *Curriculum and Evaluation Standards,* two years later the NCTM published the *Professional Standards for the Teaching of Mathematics* (NCTM 1991). It defines a teacher's major roles as

- setting goals and selecting or creating mathematical *tasks* to help students achieve these goals;
- stimulating and managing classroom *discourse* so that both the students and the teacher are clearer about what is being learned;
- creating a classroom *environment* to support teaching and learning mathematics;
- *analyzing* student learning, the mathematical tasks, and the environment in order to make ongoing instructional decisions (NCTM 1991, p. 5).

The *Professional Teaching Standards* specifies that the professional teacher of mathematics must experience good mathematics teaching, know mathematics and school mathematics, know students as learners of mathematics, and know mathematical pedagogy.

Developing Teachers' Knowledge

What teachers know is undoubtedly the most important influence on what and how they teach. The new vision of school mathematics is centered on students' investigation of worthwhile mathematical tasks. It is the selection of the task with which the teacher will engage the students in the study of mathematics that has the greatest impact on students' opportunity to learn and on their perceptions of mathematics (Lappan 1994). To select worthwhile mathematical tasks, to support classroom discourse about that task, and to assess students' understanding of the task, teachers must have knowledge about mathematics, about the teaching of mathematics, and of students as learners of mathematics.

Content knowledge

One of the most critical issues in teacher preparation is the appropriate content knowledge needed for a teacher to effectively teach mathematics in the spirit of the *Curriculum and Evaluation Standards*. Content knowledge has historically been the focus of recommendations for the improvement of the preparation of teachers. For the past seventy-five years, the belief prevailed that if the mathematics content knowledge of teachers was improved, the teaching and learning of mathematics would be improved.

However, research has failed to support a direct relationship between teachers' knowledge of mathematics and student learning (Begle 1979; Eisenberg 1977; McKnight et al. 1987). Nevertheless, few believe that teachers can teach what they do not know. Research does make it clear that teachers' knowledge of mathematics does have an impact on instruction in terms of its richness and conceptual level (Fennema and Franke 1992).

The classroom envisioned by the *Curriculum and Evaluation Standards* clearly calls for a teacher with a much deeper understanding of mathematics than that required of the teacher of seventy-five years ago. To provide significant worthwhile mathematical tasks for students, to orchestrate a discourse that allows for conjecturing and justification, to see and make connections with previously learned mathematics and other subject areas, to analyze student learning, a teacher needs to know a significant amount of mathematics. But, it is not enough to know more mathematics. Teachers must also know more about the mathematics they will teach.

Ball (1990) argues that teachers do not develop a substantive understanding of mathematics in their own elementary and secondary schooling. There is a difference between remembering and doing and understanding. Teachers' college mathematics courses, even for the mathematics major, do not afford them the opportunity to review and explicate the meanings underlying even simple content. In order for teachers to assess and push students' mathematical thinking, they must know school mathematics from a deeper perspective than that which they themselves learned in school:

> Teachers should understand the subject in sufficient depth to be able to represent the subject appropriately and in multiple ways. They need to understand the subject flexibly enough so that they can interpret and appraise students' ideas, helping them to extend and formalize intuitive understandings and challenging incorrect notions (p. 453).

What, then, constitutes adequate mathematics preparation for teachers at various levels of certification?

Both the *Professional Standards for the Teaching of Mathematics* (NCTM 1991) and a report by the MAA, *A Call for Change: Recommendations for the Mathematical Preparation of Teachers of Mathematics* (Leitzel 1991) proposed an answer to the question of the appropriate mathematical content preparation of teachers. The two reports present the same message, but it is *A Call for Change* that is the most detailed. In the report, courses are not specified, but the content to be covered in the preparation program and the number of semester hours that should be required by level of preparation is described. For all levels of preparation, prospective teachers should study topics from the broad categories of

1. number systems and number sense,
2. geometry and measurement,
3. patterns, functions, and algebra, and
4. data analysis, probability, and statistics.

For the K–4 teachers, the coverage is concrete and focused on understanding the underlying principles, meanings, and connections. For 5–8 teachers, their preparation program should incorporate and expand on the topics covered by the teachers at the K–4 level and additionally include the development of basic concepts of

5. calculus.

The mathematics preparation of secondary teachers should include content from the five categories listed above in more depth and at a greater level of sophistication than that of the K–4 and 5–8 teachers and additionally topics from

6. discrete processes, and
7. mathematical structures.

(See Leitzel [1991] for a detailed description of the specific topics to be included for each level of preparation.)

For elementary teachers (K–4) and middle-grades teachers (5–8), the content described above should be covered in a minimum of nine and fifteen semester hours, respectively, of college mathematics, with three and four years of high school college preparatory mathematics. For secondary mathematics teachers (9–12), the equivalent of a major in mathematics, including a yearlong upper-division sequence in some area of mathematics, is recommended. While these recommendations call for fewer hours than recommended by CUPM in the 1960s for elementary and junior high teachers and call for essentially the same number of hours for secondary teachers, the content coverage is broader and reflects back on the school mathematics they will teach.

Perhaps more important than the topics to be covered, the report (as its title suggests) calls for a change in the way college mathematics is taught. What prospective teachers learn is fundamentally connected with how they learn it. Hence, regardless of their level of preparation, prospective teachers should experience good mathematics teaching. They need to experience learning mathematics by doing rather than by being told. They need to use technology as an integral part of that learning to explore and solve problems.

The mathematical experiences of prospective teachers need to foster a disposition to do mathematics, the confidence to learn mathematics independently, the development and application of mathematical language and symbolism, a view of mathematics as a study of patterns and relationships, and a perspective on the nature of mathematics through a historical and cultural approach. It is in their undergraduate mathematics courses that teachers can begin to develop their knowledge of how to teach mathematics.

Pedagogical knowledge

What do prospective teachers need to know about the craft of teaching in order to implement and maintain a classroom as envisioned by the *Curriculum and Evaluation Standards?* There are general pedagogical skills that all teachers need to know. All teachers need to know how to manage a classroom, discipline unruly students, get the day or class going smoothly, keep records, meet with parents, and a perform a host of other routine activities.

Studies of novice and expert teachers indicate that one significant difference between them is that novices do not have as many appropriate mental scripts to guide them in instructional and management routines (Brown and Borko 1992). Activities that for the expert require little conscious effort place cognitive demands on the novice. Since dealing with routine activities distracts from the real job of teaching mathematics, preparation programs need to help prospective teachers develop a repertoire of appropriate scripts that will serve them as a set of survival skills when they begin to teach.

While some routine tasks might be viewed as generic and applicable to teaching any discipline, something as routine as leading a classroom discussion in mathematics is different from leading a classroom discussion in another discipline such as social studies. In mathematics more emphasis must be placed on the justification of ideas rather than their generation.

Similarly, group work is different in mathematics than in another subject area like science. In science it is the process that is important. But in mathematics, it is the negotiation of meaning that is central. Teaching mathematics is not the same as teaching reading or science or social studies. Teachers of mathematics not only need to know general classroom strategies and teaching techniques, but they also need to know methods that are specific to mathematics.

Pedagogical content knowledge

The special knowledge about teaching that is specific to the content domain being taught has been labeled *pedagogical content knowledge* (Shulman

1986). It includes a knowledge of how the subject might be represented for learners and knowledge of how students think within specific content domains. Knowledge of the curriculum and available instructional materials and technologies could also be included in pedagogical content knowledge.

Knowledge of mathematical representations is not totally separate from mathematics content knowledge. In fact, if the content preparation of teachers included an opportunity to revisit school mathematics from a deeper perspective, both the real-world situations and the concrete and pictorial representations that model school mathematics might be developed.

Unfortunately, prospective teachers can often perform mathematical computations correctly but are unable to give a real-world problem from which the computation might be derived (Ball 1990). If teachers are to orchestrate mathematical discourse, they must be able to draw on a wide range of mathematical representations and contexts for applying mathematics.

The new vision of school mathematics is learner centered. Choosing these tasks, directing the discourse in the classroom, and analyzing students' understanding requires a knowledge of how students' mathematical thinking most naturally evolves. There is a fairly robust knowledge base available on students' development of early number concepts (Carpenter and Moser 1984).

Further, it has been demonstrated that teachers empowered with a knowledge of children's thinking in this domain can significantly change their instructional practice to focus more on problem solving (Carpenter et al. 1989). Unfortunately, similar robust, integrated knowledge bases are not available for most content areas within the mathematics curriculum. Hence, the teacher must rely on his or her own questioning, observing, and listening to students.

Development of this inquiry approach to teaching must become the goal of the preparation program. The techniques of alternative assessment can play a critical role in enabling a teacher to analyze and understand students' thinking and learning. Observations of a performance task, the review of a portfolio of student work, or an interview with a student are all means for gaining knowledge of a student's mathematical thinking. Prospective teachers need information and practice in using and managing these new assessment practices.

Along with a knowledge of mathematical representations and of students, equally important is a knowledge of the mathematics curriculum robust enough to enable teachers to see the big ideas. Most prospective teachers have come through a school mathematics curriculum that consists of isolated bits and pieces, each of which seems as important as the next. If less is to be more, teachers must be able to judge what "the less" should be. Further, if teachers are to evaluate and implement instructional materials and technologies, they must do so on the basis of the extent to which the materials or computer programs embody significant mathematics for that grade level.

Teachers' content and pedagogical knowledge and their knowledge of students play critical roles in determining what goes on in the mathematics classroom. But knowledge alone is not the determining factor. Knowledge is colored by beliefs.

Developing Teachers' Beliefs

What teachers believe about the nature of mathematics and about the nature of mathematics learning and teaching has been shown to be as important as their knowledge of mathematics (Thompson 1992). The vision of school mathematics presented in the *Curriculum and Evaluation Standards* is one that views mathematics as problem solving and learning as the construction of knowledge through working on worthwhile mathematical tasks.

However, a teacher who believes that mathematics is a set of facts and procedures often is one who views the teaching of mathematics as the clear presentation of concepts and procedures. A teacher who believes knowing mathematics is the ability to recite facts and perform computations often is one who views the learning of mathematics as drill and practice.

Research has shown a fairly consistent relationship between teachers' beliefs about the nature of mathematics and their instructional practice (Thompson 1984). Similarly, a strong relationship has been observed between teachers' beliefs about teaching and their beliefs about how students learn mathematics (Cobb, Wood, and Yackel 1990). However, the relationship between teachers' beliefs about teaching and learning mathematics and their practice can be at odds.

For example, a teacher may believe in a problem-solving approach to teaching but find that the demands or expectations of the school to improve standardized test scores or to cover the district syllabus forces him or her to teach in a manner inconsistent with this belief. However, a teacher who believes that teaching is mostly telling is unlikely to forsake that belief for a problem-solving approach.

Regardless of the relationship to practice, beliefs have been shown to be resistant to change. Beliefs about the nature of mathematics and what it means to teach and to learn mathematics are developed over one's entire life as a student of mathematics. Teacher preparation programs—no matter how excellently they model the desired conceptions—make up only a small fraction of the prospective teacher's total experience of mathematics.

Recently, reflection has been identified as a necessary ingredient of change (LaBoskey 1994). One vehicle for reflection is the use of cases in teacher preparation programs in much the same manner as they are used in business or medical schools (Barnett 1991; and Merseth 1991).

Another technique used to cause the prospective teacher to examine his or her current beliefs is action research, in which the student is required to conduct a small research study during the clinical experience (Noffke and Brennan 1991). Research-based information about children's thinking in solving simple addition and subtraction word problems has been shown to be a powerful agent in changing primary teachers' beliefs and instructional practice. Whether it also can be used to change prospective teachers' beliefs is currently under investigation at a number of universities (Lubinski, Otto, and Rich 1993).

It is clear that teacher preparation programs need to address teachers' beliefs as well as teachers' knowledge. Teachers' knowledge of mathematics and pedagogy are translated into practice through the filter of their beliefs about mathematics and the nature of mathematics teaching and learning. This underscores the need for a change in the way undergraduate mathematics courses are taught.

Developing Teachers' Classroom Practice

Student teaching and other clinical experiences are viewed by teachers as the most valuable component of their preparation programs. It is through these classroom experiences that prospective teachers have the opportunity to begin to develop their practice. However, a number of studies have found that student teachers tend to adopt the behaviors and attitudes of their cooperating teachers in the course of the student teaching experience (Brown and Borko 1992). It is unclear whether this is a function of the cooperating teacher, the students in the classroom, or the general environment.

If prospective teachers are to develop the ability to create and manage engaging learning environments were students are actively involved, it must begin during student teaching. Hence there is a critical need for supportive placement sites that provide and encourage reform-based instruction in the preparation process.

To break the cycle of teaching as prospective teachers perceived that they were taught, field experiences must challenge their beliefs about teaching and learning and provide an opportunity for student teachers to reflect on their practice. Rather than offering suggestions, the college supervisor and cooperating teacher need to provoke the student teacher to analyze his or her own lessons.

Preparation programs must develop a prospective teacher's knowledge of mathematics content and pedagogy and cultivate a disposition toward mathematics and its teaching and learning that is often in sharp contrast to dispositions developed over the prospective teacher's own schooling. Further, there must be opportunities to develop classroom practice in a culture that supports good teaching and the reform vision of school mathematics. These critical issues in the preparation of teachers of mathematics are being addressed by some promising new programs.

SOME PROMISING INNOVATIONS

The innovations described below address to various degrees these critical issues. The Holmes model and the elementary/middle school mathematics specialist program focus on developing teachers' content knowledge. Programs using hypermedia platforms focus on developing teachers' pedagogical content knowledge and beliefs. Professional development schools and induction programs focus on developing practice sites and the co-reform of changing both how teachers are prepared and the environment in which they practice and teach.

The Holmes Model

In the mid-1980s, a number of research universities across the U.S. formed a coalition called the Holmes Group, which was organized around the twin goals of reform in teacher education and reform in the teaching profession. The group contends that increased subject matter competence in teacher preparation is needed along with the establishment of standards for the teaching profession.

To address the concern for increased subject matter knowledge, the Holmes model calls for prospective teachers to have an undergraduate major in a teaching field and requires a fifth year for the completion of a teacher preparation program, thus eliminating baccalaureate degree preparation programs at participating institutions (Holmes Group 1986).

The Ohio State University is one Holmes Group member to implement the Holmes model for the preparation of secondary mathematics teachers. Graduates with a major in mathematics who are interested in teaching mathematics enroll in a five-quarter program: two summers and the intervening year. In addition to the usual educational foundations and methods courses, students take a research design course in the fall quarter, design a research project in the winter quarter, and carry out the research project in the spring quarter during their student teaching.

During the summer capstone seminar, students write up their research projects, which then become their master's projects. Through the research component, the Ohio State program seeks not only to cultivate teachers as researchers but to induce an examination of beliefs through action research as described earlier.

The program further strives to develop reflective practitioners, with more attention during student teaching to planning for instruction and an examination of the relationship of planning to what happens in the classroom. Hence, this Holmes model addresses the need to increase prospective teachers' content knowledge and the development of their pedagogical content knowledge and beliefs.

The Elementary Mathematics Specialist Model

Another approach to increased subject matter competence was taken by the state of Illinois. Rather than a five-year program with an undergraduate major in a teaching field, Illinois requires all prospective elementary teachers, as part of the undergraduate general education requirement for certification, to take eighteen semester hours in a teaching field, nine of which are at the upper division level. Illinois State University has developed a program featuring a sequence of seven mathematics courses specifically designed for prospective teachers of elementary and middle school mathematics.

Unlike the undergraduate major in mathematics that begins with calculus, this program builds to calculus. The content covered is consistent with the content recommended for middle grade teachers in *A Call for Change* (Leitzel 1991). Each course seeks both to extend students' knowledge and mathematical power and to relate the content learned to the school curriculum. Since few university departments have a sequence of eighteen hours designed for prospective elementary teachers, the program in mathematics enjoys healthy enrollments. The result is a larger than usual number of elementary teachers being certified in Illinois with a substantial amount of mathematics in their background.

Use of Hypermedia Platforms

To address the need to develop prospective teachers' pedagogical content knowledge and beliefs, hypermedia platforms are being introduced in elementary education mathematics methods courses at Michigan State University, University of Michigan, and Vanderbilt University (Goldman and Barron 1990; Goldman, Barron, and Witherspoon 1991; Lampert 1994; and Lampert and Ball 1990).

Interactive, multimedia classrooms are equipped with workstations configured with a computer with a CD-ROM drive, videodisc player, video monitor, and audiocassette recorder and headphone amplifier. These hypermedia platforms provide the opportunity to combine video segments of actual classrooms with related print information such as students' work, teachers' journals, and relevant research to create information-rich cases for prospective teachers to analyze.

Students can compare two different teachers' handling of similar situations, explore children's thinking on a particular concept, or anticipate the resolution of an instructional problem. They can stop and replay the tape to look at events more closely, connect what is happening in the lesson with teachers' and students' writings about the lesson or to another related video segment, and track their own developing interpretations.

Videos of actual teaching represent the complexities involved in the act of teaching. The hypermedia cases provide prospective teachers with an opportunity to see alternatives to traditional approaches to teaching. By engaging in the analyses required, prospective teachers are forced to examine their own understanding of mathematics and their beliefs about teaching.

Professional Development Schools

One of the challenges in teacher preparation is finding school environments for prospective teachers to observe and practice that mirror the new vision of school mathematics. One attempt at addressing this concern is the identification of professional development schools. A professional development school is a working model of restructured schools developed and operated by local school and university educators working as colleagues.

These schools strive to offer exemplary K–12 programs, serve as an institutional base for the preparation of new teachers, and conduct applied research and curriculum development. The effort can be described as co-reform: cooperative working to reform preparation and practice.

Holt High School in Holt, Michigan, is a Michigan State University Professional Development School (Michigan State University 1992). There are a number of ongoing projects involving almost all disciplines within the school. In mathematics, school and university faculty are working collaboratively to improve teaching and learning for understanding.

Three pedagogical studies are being conducted. One focuses on using a conceptual orientation to instruction in the most basic mathematics offered at the high school while also maintaining computational competence. A second involves giving all students access to algebra with a course that is team-taught by a university professor using functions and computers to aid in developing understanding. A third study focuses on the use of portfolios and performance assessment in Algebra 2 as a way to help students better comprehend what it means to understand mathematics.

In addition to the three studies, a mathematics teacher from Holt High School team-teaches a section of the secondary mathematics methods course with a university professor. The course is actually taught at Holt, allowing prospective teachers access to mathematics classrooms and students. The course also emphasizes teaching for understanding with a goal of helping prospective teachers understand what it means to teach and assess for understanding.

Induction Programs

Providing appropriate environments in which prospective teachers can observe and practice is one obstacle in the preparation of teachers for implementing a new vision of school mathematics. Even if preparation programs could be entirely effective in transforming prospective teachers' knowledge, beliefs, and practice to match the one envisioned by the reform movement, how can that momentum be maintained once the new teacher is hit with the realities and pressures of that first teaching job?

Some new teachers respond to this reality shock by changing their teaching in a manner contrary to the beliefs about teaching they developed during their preparation program, by changing their attitude toward teaching and learning to match those of the culture in which they find themselves, or even by leaving the teaching profession altogether (Gray and Gray 1985). Induction programs are one attempt to reduce the shock and ease the transition from student to teaching professional.

The University of Northern Colorado is one institution that operates an induction program for first-year teachers. The program is a collaborative program between the university and neighboring school district and is unrelated to any licensure or state mandate. In it, first-year and reentry teachers, called partner teachers, are provided a support team consisting of a mentor, a full-time teacher selected and employed by the school district who is responsible for providing assistance on a daily basis; a university field consultant who makes regular, scheduled classroom observations with specific feedback and conducts the companion seminars for the partner teachers and the mentors; and the building principal.

The induction program is designed to develop reflective abilities in the beginning teacher with activities such as a professional portfolio, a professional development plan, video and audiotape self-analyses, academic seminars, and the establishment of cohort groups. Follow-up studies indicate a high degree of satisfaction with the program, with both partner teachers and mentors reporting that the program enhanced their skills and caused them to be more reflective (Jacobsen 1992).

The Northern Colorado program is one of many across the country. While the specific focus of these programs generally is not the teaching of mathematics, induction programs do offer hope that the efforts to prepare teachers for a new vision of school mathematics can be maintained through the critical first years of teaching.

CONCLUSION

After seventy-five years, programs for the preparation of teachers still consist of academic subject matter, professional education, and supervised practice. This three-pronged approach now extends to the preparation of secondary and middle school, as well as elementary school, teachers.

While the components have remained the same, their substance has not. The amount of content knowledge recommended has increased for prospective elementary teachers, and the emphasis has shifted from knowing school mathematics to understanding it. For middle school teachers more subject matter expertise is expected. The major in mathematics is still the expected norm for the prospective secondary teacher, but that major has become more robust over time.

The professional education component has become more central to preparation programs and has increasingly focused on content-specific pedagogy including the use of technology and alternative assessment techniques. Programs are now concerned not only with developing prospective teachers' knowledge, but with cultivating appropriate dispositions toward mathematics and mathematics teaching and learning.

In the past seventy-five years, the vision of school mathematics has changed from the mastery of facts and procedures by a few to the mathematical empowerment of all students. With this change in vision, the job of preparing teachers has become even more challenging. Unless novice teachers have experienced good mathematics teaching as students, seen it modeled by teachers they respect, and are placed in a culture of teaching that accepts and practices good teaching, it will be unlikely that they can implement and maintain good teaching as described in the NCTM *Curriculum and Evaluation Standards* and *Professional Teaching Standards.*

Yet it is the teacher who holds the key to reform. Whether or not the reform is realized depends on how well prepared each new teacher is to embrace and hold onto the new vision.

. . . .

BIBLIOGRAPHY

Ball, Deborah L. "The Mathematical Understandings That Prospective Teachers Bring to Teachers Education." *The Elementary School Journal* 90 (1990): 449–66.

Barnett, C. "Building a Case-Based Curriculum to Enhance the Pedagogical Content Knowledge of Mathematics Teachers." *Journal of Teacher Education* 42 (2) (1991): 1–9.

Begle, Edward G. "Critical Variables in Mathematics Education: Findings from a Survey of Empirical Literature." Washington, D.C.: Mathematical Association of America and the National Council of Teachers of Mathematics, 1979.

Brown, Catherine A., and Hilda Borko. "Becoming a Mathematics Teacher." In *Handbook of Research on Mathematics Teaching and Learning,* edited by D. A. Grouws, pp. 209–39. New York: Macmillan, 1992.

Carpenter, Thomas P., and James M. Moser. "The Acquisition of Addition and Subtraction Concepts in Grades One through Three." *Journal for Research in Mathematics Education* 15 (1984), 179–202.

Carpenter, Thomas P., Elizabeth Fennema, P. L. Peterson, C. P. Chiang, and Megan Loef. "Using Knowledge of Children's Mathematics Thinking in Classroom Teaching: An Experimental Study." *American Educational Research Journal* 26 (1989): 499–532.

Cobb, Paul, T. Wood, and Erna Yackel. "Classrooms as Learning Environments for Teachers and Researchers." In *Constructivist Views on the Teaching and Learning of Mathematics, Journal for Research in Mathematics Education* Monograph No. 4, edited by Robert B. Davis, Carolyn A. Maher, and Nel Noddings, pp. 125–46. Reston, Va.: National Council of Teachers of Mathematics, 1990.

Committee on the Undergraduate Program in Mathematics. *Recommendations for the Training of Teachers.* Washington, D.C.: Mathematical Association of America, 1961, revised 1966.

Dossey, John A. "The Current Status of Preservice Elementary Teacher-Education Programs." *Arithmetic Teacher* 29 (1) (1981): 24–6.

Dubish, R. "Teacher Education." In *Mathematics Education: The Sixty-ninth Yearbook of the National Society for the Study of Education,* edited by E. G. Begle, pp. 285–310. Chicago: NSEE, 1970.

Eisenberg, Theodore A. "Begle Revisited: Teacher Knowledge and Student Achievement in Algebra." *Journal for Research in Mathematics Education* 8 (1977): 216–22.

Fennema, Elizabeth, and Megan L. Franke. "Teachers' Knowledge and Its Impact." In *Handbook of Research on Mathematics Teaching and Learning,* edited by Douglas A. Grouws, pp. 147–64. New York: Macmillan, 1992.

Gibb, E. Glenadine, Houston T. Karnes, and F. Lynwood Wren. "The Education of Teachers of Mathematics. In *A History of Mathematics Education in the United States and Canada,* Thirty-second Yearbook of the National Council of Teachers of Mathematics, pp. 301–50. Washington, D.C.: The Council, 1970.

Goldman, E., and L. Barron. "Using Hypermedia to Improve the Preparation of Elementary Teachers." *Journal of Teacher Education* 41 (3) (1990): 21–31.

Goldman, E., L. Barron, and M. L. Witherspoon. "Hypermedia Cases in Teacher Education: A Context for Understanding Research on the Teaching and Learning of Mathematics." *Action in Teacher Education* 13 (1) (1991): 28–36.

Gray, W. A., and M. M. Gray. "Synthesis of Research on Mentoring Beginning Teachers." *Educational Leadership* 43 (3) (1985): 37–43.

Holmes Group. *Tomorrow's Teachers*. East Lansing, Mich.: Holmes Group, 1986.

Jacobsen, M. "Mentoring as a University/Public School Partnership." In *Teacher Induction and Mentoring: School-Based Collaborative Programs*, edited by G. P. DeBolt, pp. 139–66. Albany, N.Y.: State University of New York Press, 1992.

LaBoskey, V. K. *Development of Reflective Practice: A Study of Preservice Teachers*. New York: Teachers College Press, 1994.

Lampert, M. L. "Why Use Interactive Media in Teacher Preparation?" *Innovator* 25 (1) (1994): 4–6.

Lampert, M. L., and Deborah L. Ball. "Using Hypermedia Technology to Support a New Pedagogy of Teacher Education." Issue paper 90–5. East Lansing, Mich.: Michigan State University, National Center for Research on Teacher Education, 1990.

Lappan, Glenda. "Professionalism Linking Teachers to Teachers: Rights, Responsibilities, and Growth." Paper presented at the annual meeting of the National Council of Teachers of Mathematics, Indianapolis, April 1994.

Leitzel, J. R. C., ed. *A Call for Change: Recommendations for the Mathematical Preparation of Teachers of Mathematics*. Washington, D.C.: Mathematical Association of America, 1991.

Lubinski, C. A., A. D. Otto, and B. S. Rich. "Collaboration among University Faculty, Experienced Teachers and Inexperienced Teachers: A Model for Teacher Preparation." Paper presented at the meeting of the National Council of Supervisors of Mathematics, Seattle, March 1993.

Mathematical Association of America. "The Reorganization of Mathematics in Secondary Education." A report by the National Committee on Mathematical Requirements. Oberlin, Ohio: MAA, 1923.

_____. "Report on the Training of Teachers of Mathematics." Report of the Commission on the Training and Utilization of Advanced Students in Mathematics. *American Mathematical Monthly* 42 (1935): 263–77.

McKnight, Curtis C., F. Joe Crosswhite, John A. Dossey, Edward Kifer, Jane O. Swafford, Kenneth J. Travers, and Thomas J. Cooney. *The Underachieving Curriculum: Assessing U.S. School Mathematics from an International Perspective*. Champaign, Ill.: Stipes Publishing Co., 1987.

Merseth, K. K. *The Case for Cases in Teacher Education.* Washington, D.C.: American Association of Colleges for Teacher Education and American Association for Higher Education, 1991.

Michigan State University. *1992–93 Plan for School and University Alliance and Educational Extension Service: 1992–93 State of Michigan Grant to Michigan State University for the Michigan Partnership for New Education,* Vols. 1–2. East Lansing, Mich.: Michigan State University, 1992.

National Commission on Excellence in Education. *A Nation at Risk: The Imperative for Educational Reform.* Washington, D.C.: U.S. Government Printing Office, 1983.

National Council of Teachers of Mathematics. *Curriculum and Evaluation Standards for School Mathematics.* Reston, Va.: The Council, 1989.

_____. "First Report of the Commission on Post-War Plans." *Mathematics Teacher* 37 (1944): 225–32.

_____. "The Place of Mathematics in Secondary Education." Final report of the Joint Commission of the Mathematical Association of America and the National Council of Teachers of Mathematics (Fifteenth Yearbook). New York: The Council, 1940.

_____. *Professional Standards for Teaching Mathematics.* Reston, Va.: The Council, 1991.

_____. "Second Report of the Commission on Post-War Plans." *Mathematics Teacher* 38 (1945): 195–221.

Noffke, S. E., and M. Brennan. "Student Teachers Use Action Research: Issues and Examples." In *Issues and Practices in Inquiry-Oriented Teacher Education,* edited by B. R. Tabachnick and K. M. Zeichner, pp. 186–201. New York: Falmer Press, 1991.

Shulman, L. S. "Those Who Understand: Knowledge Growth in Teaching." *Educational Researcher* 15 (2) (1986): 4–14.

Thompson, A. G. "The Relationship of Teachers' Conceptions of Mathematics Teaching to Instructional Practice." *Educational Studies in Mathematics* 15 (1984): 105–27.

_____. "Teachers' Beliefs and Conceptions: A Synthesis of the Research." In *Handbook of Research on Mathematics Teaching and Learning,* edited by Douglas A. Grouws, pp. 127–46. New York: Macmillan, 1992.

Who Will Teach?

Shirley M. McBay

I N KEEPING with the wishes of "the American people," the Congress passed the largest crime bill in the country's history, a bill that was signed into law by President Bill Clinton on 13 September 1994. Key provisions of the law include grants to localities to hire 100 000 officers for community policing, college scholarships for students who agree to serve as members of a "police corps," state grants to build more prisons to house violent criminals and establish "boot camps" for young offenders, and life in prison for three-time felons.

While addressing crime is clearly important to our quality of life, investing in our educational infrastructure is also essential to America's future. However, it is not clear that we could, for example, obtain broad-based and bipartisan support for grants to our poorest urban and rural school districts to hire 100 000 *teachers*. Nor is large-scale support evident for college scholarships and grants for a "mathematics *and science teacher corps*" to address the severe minority teacher shortage in most large urban school districts and to make it possible for students in these districts to be the best in the world in mathematics and science.

Where are the funds to help states and communities *build and wire schools* in their poorest districts so that students and teachers can have access to the information superhighway? Is the nation's commitment to equity and fairness deep enough to *build our human infrastructure* through the creation of positive alternatives for children and youth in desperate need of supportive living *and* learning environments?

Such investment *before* myriad problems converge can produce positive results for the thousands of low-income families who are trapped in circumstances that contribute to criminal behavior. Now is the time to ask if America

Shirley McBay is president of the Quality Education for Minorities (QEM) Network, a nonprofit organization in Washington, D.C. She has served as a member of the MIT administration, the program staff of the National Science Foundation, and the faculty of Spelman College.

has the will and commitment to fulfill its promise of equal opportunity for all. Now is the time to put aside stereotypes and myths about "innate ability" that are blinding us to clear and practical remedies that are in both our short- and long-term best interest.

If we are insightful enough to invest in education and discard negative stereotypes *now*, we should consider using an integrated service delivery approach to the development of our human resource base, linking health and social services to education and training (QEM Network, 1993). Building human infrastructure requires a deliberate, collaborative effort and a deep level of trust among service providers and recipients, including social workers, healthcare providers, community and religious leaders, teachers, parents, and students. It also requires a concerted effort to develop leadership skills within each of these groups, but especially among teachers, parents, and students.

While long-term success requires each of these groups to play multiple roles, including that of teacher, this paper focuses upon K–12 public school teachers, particularly teachers of mathematics. It seeks to answer the question of how to identify and prepare future teachers of mathematics who will have the knowledge, skills, concern, and dedication to help our youth grow and develop into productive, lifelong learners and leaders, instead of three-time losers serving mandatory lifetime sentences. Who will be our teachers? How will they be attracted, prepared, and retained in the teaching profession? *America's future may very well hinge on the answers to these questions.*

BACKGROUND

If enrollment trends in teacher preparation programs continue to remain low and teacher retirements mushroom as anticipated, we can expect a major teacher shortfall by the year 2000. Most affected by this shortage will be areas that continue to have the greatest need for quality teachers: large urban school districts and rural communities, and subject areas such as mathematics, chemistry, and physics. Exacerbating this shortage is the survey finding that *most new teachers apparently do not want to teach in schools with the greatest needs.* For example, recent survey results reported in *Teacher Magazine* indicate that 75 percent of new teachers want to teach in middle-class schools in contrast to 14 percent who want to teach in culturally diverse schools and 6 percent in low-income schools.

The significant predicted shortfall of minority teachers further amplifies the problem. Researchers have found that a series of educational, social, economic, and demographic trends have converged to create the current shortage. It is estimated that by the year 2000, minority teachers will represent only 5 percent of the national teaching force, while the student population will be one-third minority.

The recruitment, preparation, and retention of minority teachers is critical for the growth and development of both minority students and nonminority students. The quality of education for minority and nonminority students would be greatly enhanced by a bilingual, racially and ethnically mixed teaching staff pursuing the goal of academic excellence for *all* students.

Issues significant to the recruitment and retention of minorities in the teaching profession include—

- improving the academic preparation of students from prekindergarten through graduate school,
- improving preservice and in-service teaching programs,
- improving professional development for faculty and staff at institutions of higher education,
- developing comprehensive marketing and recruitment strategies to help persuade well-prepared college graduates to enter the teaching profession,
- targeting specific groups from which to identify interested, talented individuals as future teachers (e.g., residents of low-income public housing), and
- improving the image of the teaching profession.

The recruitment and retention of minorities into the teaching profession must become a national priority, for as table 1 shows, a major mismatch exists between the racial and ethnic composition of the public school student population and the public school teacher workforce. For instance, African American students in 1987–88 constituted more than 16 percent of public school students, while only eight percent of the K–12 teacher workforce was African American.

Low representation of minorities among teachers limits opportunities for minority and other children to see minority adults in leadership positions. The Council of Great City Schools reports that in thirty-five of the forty-five Great City School Districts, minority student enrollment is higher than nonminority enrollment (Great City Schools enroll 12 percent of all students and employ 12 percent of all teachers). Overall, enrollment in the forty-five districts is approximately 70 percent minority, whereas the teacher workforce in these districts is 68 percent white, essentially the reverse. Minority teachers provide important role models for students and facilitate relationships between school and home and the community.

Not only is minority representation in the overall teacher work force low, it is worse among teachers of mathematics and science (see Table 3 below). The highest representation for blacks is among special education teachers while the lowest for Hispanics is mathematics and science.

· ·

Table 1. Percentage Distribution of Public School Teachers and Students by Race-Ethnicity: 1987–88

Group	American Indian	Asian	Black, non-Hispanic	Hispanic	White, non-Hispanic
Students	1.1%	2.5%	16.3%	9.4%	70.7%
Teachers	1.0%	0.9%	8.0%	2.6%	87.5%

Source: *America's Teachers: Profile of a Profession*, U.S. Department of Education, Office of Educational Research and Improvement, National Center for Education Statistics, from figure 2.3, page 20, NCES 93-025

· ·

Table 2. Size/Percent Total Public School Teacher Workforce, by Community Type, 1987–88

Total	Urban	Suburban	Rural/Small City
2,093,637	549,875	451,457	1,092,305
% of Total Workforce	26.3%	21.6%	52.2%

Source: *America's Teachers: Profile of a Profession*, U. S. Department of Education, Office of Educational Research and Improvement, National Center for Education Statistics, Abstracted from table 2.7, page 12, NCES 93-025

· ·

Table 3. Percentage Representation of Public School Teachers in Selected Disciplines, by Race and Ethnicity, 1987–88

Group	American Indian	Asian/Pacific Islander	Black non-Hispanic	Hispanic	White, non-Hispanic
% Total Teacher Workforce	1.0%	0.9%	8.0%	2.6%	87.5%
K–Elementary	0.9%	1.1%	8.8%	2.6%	86.5%
Math/Science	0.9%	1.1%	7.4%	1.8%	88.7%
Special Education	1.3%	0.8%	9.1%	2.0%	86.8%
Bilingual/ESL	0.8%	5.3%	5.4%	36.4%	52.1%
Social Studies	0.9%	0.7%	6.4%	2.6%	89.4%

Source: *America's Teachers: Profile of a Profession*, U. S. Department of Education, Office of Educational Research and Improvement, National Center for Education Statistics, Abstracted from table 2.10, page 15, NCES 93-025

· ·

Later tables provide information on the number and percent of public school teachers by race, gender, and community type, 1987–88 (table 4) as well as on black public school teachers by gender and community type, 1987–88 (table 5).

The need to encourage more minorities to enter the teaching profession is clear. By the year 2000, the minority teaching force is projected to represent only 5 percent of the national teaching force at a time when minority students will comprise at least one-third of all students. In addition, of the 700 000 new teachers being trained, only about 35 000 (or 5 percent) are expected to be minority. Of the 100 000 minority students graduating from colleges and universities annually, only about 10 000 major in elementary and secondary education. Nevertheless, it is important to keep in mind that minority teachers historically have tended to be first-generation college students: This has implications for potential sources of new teachers.

Socioeconomic status is a major determinant of access to quality mathematics and science instruction and advanced courses. For example, the correlation between enrollment in advanced mathematics and science courses and socioeconomic status has been clearly established (see table 6). We know as well that advanced offerings in mathematics and science at schools decrease as the percentage of enrolled minority students increases.

Linked to the paucity of advanced offerings is the relatively low percentage of certified mathematics and science teachers in schools that serve low-income minority students. The National Science Board in its 1991 *Science and Engineering Indicators* cited findings that the percentage of certified mathematics and science teachers decreases as the percentage of enrolled minority students increases.

- In 1987, only 39 percent of the secondary school teachers who taught low-ability classes in low SES (socioeconomic status), minority, inner-city schools were certified to teach science or mathematics at the secondary level, compared with 82 percent of the teachers of low-ability classes in high-wealth, predominantly white, suburban schools.
- Low-track students in the most advantaged schools (high SES, white, suburban) were more likely to have better qualified teachers of science and mathematics than high-track students in the least advantaged schools (low SES, high-minority, inner-city).

Not surprisingly then, students in low-income communities, including low-income public housing, are less likely to have access to advanced mathematics and science courses or to have certified mathematics and science teachers because they are more likely to be enrolled in predominantly minority schools.

Well-prepared minority teachers of mathematics and science are especially needed to address the crisis in educational quality existing in urban communities, particularly in predominantly minority schools. Any strategy to significantly increase the number of talented minority students interested in pursuing mathematics teaching careers must address trends that have contributed to the shortage.

These trends include the improvement in educational and occupational opportunities that is leading minority graduates away from teaching towards better paying jobs; the persistent low prestige accorded to the teaching profession relative to other professions; the dwindling availability of financial aid for low- and middle-income students (Darling-Hammond 1987); and the disproportionately high failure rates of minority teachers on teaching certification examinations (Smith 1989). This last trend has weakened the credibility and quality of teacher preparation programs at predominantly minority institutions, long the dominant source of minority teachers (Hatton 1988).

A related trend has been the attrition of experienced minority teachers. The *1988 Metropolitan Life Survey of the American Teacher* found that 40 percent of the minority teachers surveyed said they were likely to leave the profession in the next five years, compared to 25 percent of nonminority teachers. Further, Smith (1989) estimates that between 25 and 30 percent of current minority teachers, particularly minority teachers in the South, are rapidly approaching retirement.

SOME CRITICAL ISSUES

Precollege and Undergraduate Levels

Major barriers at the precollege and undergraduate levels often derail efforts to significantly increase either the number of undergraduate minority mathematics majors who complete the bachelor's degree in mathematics and seek a master's in teaching, or the number of minority mathematics students who enroll in teacher preparation programs and begin teaching immediately following graduation. Some of these issues are presented below.

Precollege Academic Enrichment and Guidance

- Lack of quality precollege mathematics instruction (e.g., curriculum, teacher preparation, technology, and hands-on experiences)
- Lack of interest and motivation among minority students toward mathematics beginning as early as grade four
- The effects of low expectations

Some Barriers to Effective Recruitment at the Precollege/ Undergraduate Levels

- Inadequate information on mathematics teacher education programs
- Lack of clearly defined mathematics education teacher preparation programs, particularly at predominantly minority institutions
- Lack of financial aid specifically for students interested in pursuing mathematics teaching careers
- Absence of specific programs and financial assistance for talented minority students enrolled in two-year colleges who could pursue teacher preparation programs in mathematics at four-year institutions
- Lack of emphasis on excellence in teaching at many colleges and universities and the absence of incentives for faculty who may have an interest in precollege education to get involved
- The low status of teacher preparation programs on college campuses

Some Barriers to Effective Teacher Preparation at the Undergraduate Level

- Inadequate infrastructure at many predominantly minority institutions to offer quality mathematics education programs
- Inability of many graduates to meet professional and certification standards
- Small number of mathematics faculty at predominantly minority institutions actually involved in teacher preparation programs
- Inability of many minority teacher education graduates to pass state teacher examinations and to deal effectively with the realities of rural and urban schools
- Administrative barriers to establishing or strengthening of mathematics teacher preparation programs

Graduate Programs and Teacher Professional Development

A number of barriers exist at institutions that are the major producers of minority teachers. These barriers prevent the production of a significant number of minority mathematics graduates who pursue graduate degrees in teaching. Also, they limit the nature of the professional and leadership development opportunities available to K–12 teachers of mathematics through these institutions.

For students to receive the quality and level of instruction they need to be mathematically literate or to successfully pursue a career as a mathematics teacher, it is critical that their mathematics teachers understand fundamental mathematical concepts and are prepared to teach these concepts in ways that motivate students to pursue increasingly challenging mathematics courses. This means teachers must have opportunities to become familiar with new curriculum materials and effective strategies, and must be prepared to meet increasingly stringent professional standards.

Not only must teachers of mathematics be comfortable and confident teaching mathematical concepts and principles, they must hold high expectations of *all* of their students to succeed. This requires greater sensitivity to, and understanding of, the cultural and linguistic diversity as well as the socioeconomic backgrounds of the students in their classrooms.

Even in the cases of predominantly minority institutions and nonminority institutions that are known to have quality advanced degree, teacher enhancement, and certification programs, these institutions generally have not taken steps to assist other institutions in strengthening and increasing the visibility of their teacher education programs.

Barriers to Teacher Leadership Development

Too often, teachers feel isolated from their peers and are unable to link research findings to their work or to have *their* experiences drive the education research agenda. Many are doubtful about the effectiveness of policies imposed from the outside. Hence, it is important to articulate ways to introduce teachers to the national picture while giving added value and credence to what is required to make these policies work in their schools and in their individual classrooms.

Teachers must be an integral part of the education reform movement as it focuses attention on the establishment of professional standards for teachers, on subject content and opportunity-to-learn standards, and on salary and working conditions. They must be included in the dialogue as policymakers move to address the growing crisis concerning the quality and number of teachers available in our public schools.

Classroom teachers are increasingly inundated with new recommendations for pedagogy, often without understanding the context for the changes or feeling comfortable with what these ideas mean in practice. The nation's schools are clearly in the midst of a transition period in which reform proposals are trickling down from administrators, policymakers, researchers, and subject-area organizations without fully being integrated into practice.

Many teachers report being overwhelmed by new state mandates and requirements aimed at revitalizing their disciplines. Attempts are being made to introduce new teaching approaches without teachers' input. Teachers, particularly minority teachers, are not playing significant roles in shaping these changes at any level.

Teachers need opportunities to hone their leadership skills and abilities and to use them to inform and empower peers to take an active part in mathematics education reform. Targeted efforts are needed to strengthen teachers' public speaking and media skills, along with their technical knowledge, so that they can serve as effective spokespersons on the national, state, and local levels for the reform of mathematics education. As such, they can help to ensure that those involved in reform efforts are aware of, and responsive to, the implications of their efforts for the mathematics education of low-income and minority children and youth.

Teachers of mathematics need regional and national exposure as well as to be identified as advocates and catalysts for change. They need to have linkages to teachers in other parts of the K–16 mathematics education pipeline, especially the segments immediately before and after where their primary mathematics teaching responsibilities lie. They also need access to information and databases on successful mathematics strategies at the K–12 level that are in place.

To ensure that teachers have such opportunities and resources, institutions of higher education need to modernize their teacher preparation and enhancement programs and make developing strong working relationships with K–12 schools in the local community a major priority.

Access to Advanced Technologies

There is a critical need to incorporate advanced technologies into teacher preparation and teacher enhancement programs so that teachers are more knowledgeable of, and comfortable with, using a range of technologies in their classrooms. Teachers need to become familiar with the potential of various technologies to provide

(a) visual images of various concepts in mathematics,
(b) simulation opportunities that break difficult concepts into manageable chunks,
(c) access for themselves and their students to various information data bases, including curriculum data bases,
(d) nonthreatening opportunities for students and teachers to review concepts that are not thoroughly understood, and
(e) computer networking opportunities with peers to facilitate exchange of information on effective teaching strategies and curriculum materials.

Opportunities are needed for teachers to have hands-on experiences with electronic bulletin boards; computer networking; interactive videodiscs; and learning via satellite, e-mail, teleconferencing, and other forms of telecommunications. This experience can enhance their teaching knowledge and effectiveness as well as their ability to motivate their students to be more successful in mathematics.

SOME STRATEGIES TO ACHIEVE TEACHER DIVERSITY AND QUALITY

To provide a school environment that supports academic success for every child requires a racially and ethnically diverse teacher workforce of the highest quality. To achieve such a workforce requires a multiple-entry pathway into the teaching profession as well as making teaching an attractive option for talented people to once again pursue.

Nontraditional sources and routes to identify and develop quality teachers must also be used, including

- residents of low-income public housing,
- families receiving public assistance,
- students enrolled in community colleges,
- retired career scientists and engineers, and military personnel, and
- a targeted effort to attract more minority males as teachers.

To attract and prepare the diverse, high-quality teacher workforce we seek, major efforts are needed that lead to

- early identification and establishment of bridge programs,
- the use of the media and other marketing strategies,
- institutional collaborations,
- a restructuring of teacher preparation programs,
- innovative professional development opportunities, and
- an enhanced image of the teaching profession.

Potential Strategies at the Precollege Level

Strategies are needed for increasing the number of minority students entering and remaining in the mathematics teacher pipeline, beginning as early as the middle school level. A continuum of programs and activities is needed for minority precollege students aimed at stimulating and preparing them to become successful mathematics majors in college and at increasing their interest in teaching as a career. (See appendix, tables 7–9, for information on SAT scores and intended undergraduate majors of college-bound seniors, and interest in teaching as a career among college freshmen.)

Potential strategies at the precollege level include the following:

- Partnerships between predominantly minority institutions and predominantly minority school districts to provide peer and cross-age tutoring opportunities in mathematics for precollege students
- Partnerships that provide summer hands-on mathematics opportunities for talented precollege students
- Precollege intervention programs in mathematics, such as Saturday math academies; summer math and computer camps for elementary and middle school students that emphasize problem-solving, cross-age and peer tutoring, team projects, and the use of technology in these activities; future teacher clubs and teacher cadet programs; and summer math institutes on college campuses for high school students that incorporate a community service component of peer and cross-age tutoring
- Activities to ensure that guidance counselors are aware of what is required for mathematics teaching careers
- Mentoring programs for high school students involving college mathematics students and faculty
- Apprenticeships for high school students with outstanding college mathematics professors
- Activities to inform minority precollege students of careers in mathematics teaching, including informal mathematics education and apprenticeship programs developed by museums, churches, community-based organizations, and industry; and the use of media

Potential Strategies at the Undergraduate Level

Strategies are needed that encourage undergraduate mathematics majors to enroll in teacher preparation programs as well as graduate school so that they can become certified as lead teachers in their fields. Examples include programs that encourage and prepare minority students in two-year community colleges to enter mathematics teacher education programs at four-year institutions, forgiving student loans for those who enter the teaching profession and work in communities with the greatest needs, and providing strong assistance with job placement upon graduation.

Existing mathematics teacher preparation programs at predominantly minority institutions need to be reviewed and a comprehensive strategy developed for adding new programs or strengthening existing ones by institutional type (two-year, four-year/university).

This strategy should apply to predominantly minority institutions that already have significant mathematics education resources and faculty; to institutions that, with additional resources, could develop strong mathematics teacher preparation and enhancement programs; and to institutions that could initiate collaborative programs with institutions that have strong mathematics teacher preparation and enhancement programs.

Examples of possible efforts include the following:

* Collaborations among predominantly minority institutions (PMIs) and between minority and nonminority institutions to facilitate the exchange of information and the sharing of resources
* Networks that will link faculty across institutions so they will remain current on teacher preparation program strategies and educational research findings
* Collaborations between four-year PMIs and two-year colleges to strengthen mathematics programs at the latter and to facilitate student transfer between two-year and four-year institutions into mathematics teacher preparation programs
* Assessment and planning strategies to assist PMIs in ascertaining the strengths and weaknesses in their teacher education programs and to develop long-term improvement plans

Potential Strategies for Professional and Leadership Development

Strategies are needed to ensure that teachers understand the major initiatives focused on mathematics education reform so that they can speak confidently about these efforts and accompanying issues with policymakers as well as with peers. Efforts are needed to further develop the leadership potential, and broaden the experience, of teachers of mathematics so that they can confidently and competently address *national* education policy issues such as

* the national education goals,
* national standards in mathematics,
* changing demographic patterns and their implications for an economy increasingly based on science and technology,
* national professional teaching standards,
* national tests and other emerging methods of assessment,
* multicultural education and inclusive curricula,
* school and teacher accountability, and
* how to effectively respond to the realities in most urban and rural schools.

Teachers must be able to address *state and local* education issues such as

* content enhancement and curriculum development in mathematics,
* effective pedagogy in mathematics at the K–12 level,
* use of advanced technologies in the classroom, and
* out-of-school mathematics education opportunities.

Strategies should be explored that enable mathematics teachers who, in partnership with mathematics faculty at nearby PMIs, could develop and sustain professional and leadership development opportunities for other teachers. Possibilities include

- curriculum groups of college and precollege teachers,
- mathematics teachers electronic networks,
- newsletters,
- regional conferences,
- state or regional mathematics teacher resource centers,
- mathematics-based community outreach projects, and
- use of the media.

Potential Sources of Teachers

Significant nontraditional sources for prospective minority teachers exist, including the military, the business community, community colleges, national laboratories, teachers' aides, docents and guides in science museums and centers, and former college students and graduates who may want to reenter the work force after taking time off for family or other reasons. Low-income public housing and low-income communities represent untapped, potential sources of talented individuals with an interest in teaching who, with additional formal training and hands-on experience, could become excellent teachers.

FINAL WORD

A national effort is needed aimed at publicizing the critical national need for significant numbers of talented minorities to enter the teaching profession. However, the locus of efforts to recruit and retain minority teachers lies at the state and local levels. Awareness, understanding, and support of minority teachers on a national level are critical to the enhancement of the education of minority students. To be successful, local communities must be the genesis of specific goals and plans for the recruitment of minority teachers for their schools.

The same sense of urgency that creates crime legislation must now be channeled to the building of our human infrastructure through a diverse teacher workforce of the highest quality.

ADDITIONAL SUPPORTING TABLES

Table 4. Number/Percent of Public School Teachers by Race, Gender, and Community Type: 1987–88

Urban Areas	Male	Female	Total	% of Total
White	121,786	297,246	419,032	76.2%
Black	18,165	70,864	89,029	16.2%
Other	11,285	30,529	41,814	7.6%
Total	151,236	398,639	549,875	100.0%
% of Total	27.5%	72.5%	100.0%	

Suburban Areas	Male	Female	Total	% of Total
White	129,260	280,923	410,183	90.9%
Black	4,815	19,261	24,076	5.3%
Other	4,237	12,961	17,198	3.8%
Total	138,312	313,145	451,457	100.0%
% of Total	30.6%	69.4%	100.0%	

Rural/Small City	Male	Female	Total	% of Total
White	299,831	706,095	1,005,926	92.1%
Black	10,926	40,403	51,379	4.7%
Other	10,662	24,338	35,000	3.2%
Total	321,419	770,886	1,092,305	100.0%
% of Total	29.4%	70.6%	100.0%	

Source: *America's Teachers: Profile of a Profession,* U.S. Department of Education, Office of Educational Research and Improvement, National Center for Education Statistics, Abstracted from table 2.7, page 12, NCES 93-025

· ·

Table 5. Black Public School Teachers by Gender and Community Type, 1987–88

Gender	Male	Female	Total
Number	33,906	130,578	164,484
% of Total	20.6%	79.4%	100.0%

Distribution by Community Type	Urban	Suburban	Rural/Small City
100.0%	54.1%	14.6%	31.2%

Source: *America's Teachers: Profile of a Profession,* U.S. Department of Education, Office of Educational Research and Improvement, National Center for Education Statistics, Abstracted from table 2.7, page 12, NCES 93-025

Table 6. Percentage of College-Bound Seniors Who Took Natural Sciences or Mathematics in High School and Average Years of Mathematics/Science Taken, by Racial/Ethnic Group and Gender: 1991

Subject/Group	Geometry	Trig.	Calculus	Honors Math	Physics	Honors Science	Yrs. of Math	Yrs. of Science
American Indian	89%	45%	11%	14%	33%	13%	3.6	3.1
Asian/Pac. Isl.	94%	72%	38%	37%	64%	32%	3.9	3.4
Black	86%	43%	9%	13%	32%	13%	3.6	3.0
Mexican Amer.	93%	44%	13%	21%	34%	19%	3.6	3.0
Puerto Rican	89%	51%	10%	15%	42%	14%	3.6	3.1
White	94%	56%	19%	24%	44%	23%	3.8	3.3
Gender								
Male	93%	58%	22%	24%	51%	22%	3.8	3.3
Female	92%	53%	17%	22%	37%	21%	3.7	3.2

Source: *Women and Minorities in Science and Engineering: An Update*, National Science Foundation, January 1992, Abstracted from table 29, page 126, NSF 92-303

• •

Table 7. American Freshmen Indicating Elementary or Secondary School Teacher as Career Choice, by Racial/Ethnic Group and by Gender: 1980 and 1990

Group/ Year	Amer. Indian	Asian/ Pacific Isl.	Black	Hispanic	White	Total	Male	Female	S/E Freshmen
1980	4.8%	1.3%	4.3%	3.7%	6.8%	8.8%	4.0%	13.2%	0.7%
1990	9.5%	2.4%	6.7%	7.8%	10.5%	12.4%	5.6%	17.9%	1.8%

Source: *Women and Minorities in Science and Engineering: An Update*, National Science Foundation, January 1992, Abstracted from table 36, page 134, NSF 92-303

• •

Table 8. Mean SAT Scores by Race/Ethnicity and Gender for College-Bound Seniors in 1981 and (1991)

Score/ Group	SAT-V	SAT-M	Total	SAT-V ≥650	SAT-M ≥650
American Indian	391 (393)	425 (437)	816 (830)	1%	3%
Asian/Pac. Isl.	397 (411)	513 (530)	910 (941)	5%	22%
Black	332 (351)	362 (385)	694 (736)	< 1%	1%
Mexican American	373 (377)	415 (427)	788 (804)	1%	3%
Puerto Rican	361 (361)	396 (406)	757 (767)	1%	2%
White	442 (441)	483 (489)	925 (930)	3%	10%
Gender					
Male	430 (426)	492 (497)	922 (923)	3%	14%
Female	418 (418)	443 (453)	861 (871)	3%	6%

Source: *Women and Minorities in Science and Engineering: An Update*, National Science Foundation, January 1992, Abstracted from table 36, pages 127 and 128, NSF 92-303

Table 9. Intended Undergraduate Majors and Corresponding SAT Mathematics Score of College-Bound Seniors by Field, Racial/Ethnic Group, and Gender: 1991

Major (%)	Education	Mathematics	Physical Science	Enginneering
American Indian	(8%) 416	(1%) 575	(1%) 527	(9%) 503
Asian/Pac. Isl.	(3%) 468	(1%) 630	(2%) 620	(17%) 581
Black	(5%) 360	(—) 489	(1%) 457	(11%) 442
Mexican American	(7%) 398	(1%) 530	(1%) 496	(12%) 490
Puerto Rican	(5%) 372	(—) 547	(1%) 482	(12%) 459
White	(9%) 450	(1%) 616	(2%) 577	(10%) 569
Gender				
Male	(4%) 454	(1%) 623	(2%) 587	(18%) 550
Female	(12%) 437	(1%) 585	(1%) 541	(4%) 539

Source: *Women and Minorities in Science and Engineering: An Update,* National Science Foundation, January 1992, Abstracted from table 34, page 131

• •

BIBLIOGRAPHY

American Council on Education. *One Third of a Nation: A Report of the Commission on Minority Participation in Education and American Life.* Washington, D.C.: American Council on Education and the Education Commission of the States, 1988.

Choy, Susan P. (MPR Associates, Inc.), and Sharon A. Bobbitt (National Center for Education Statistics). *America's Teachers: Profile of a Profession.* NCES 93–025. Washington, D.C.: U.S. Department of Education, Office of Educational Research and Improvement, National Center for Education Statistics, 1993.

Council of the Great City Schools. *Teaching and Leading in the Great City Schools,* Washington, D.C.: The Council, 1988.

Darling-Hammond, Linda, K. J. Pittman, and C. Ottinger. "Career Choices for Minorities: Who Will Teach?" Washington, D.C.: Paper prepared for the National Education Association and the Council of Chief State School Officers, 1987.

Hatton, Barbara. "Game Plan for Ending the Minority Teacher Shortage." In *Minority Access.* Washington, D.C.: National Education Association, 1988.

The Metropolitan Life Survey of The American Teacher, 1988: Strengthening the Relationship between Teachers and Students, Conducted for Metropolitan Life Insurance Company by Louis Harris and Associates, Inc. Fieldwork: April–June 1988. New York: Louis Harris and Associates, Inc.

National Science Board. *Science and Engineering Indicators, 1991.* 10th ed. National Science Board NSB 91-1. Washington, D.C.: U.S. Government Printing Office, 1991.

Quality Education for Minorities Network. *Opening Unlocked Doors: A National Agenda for Ensuring Quality Education for Children and Youth in Low-Income Public Housing and Other Low-Income Residential Communities.* Washington, D.C.: Quality Education for Minorities Network, May 1993.

Smith, G. Pritchy. *Increasing the Number of Minority Teachers: Recommendations for a Call to Action.* Cambridge, Mass.: The Quality Education for Minorities Project, Massachusetts Institute of Technology, March 1989.

White, Patricia E. *Women and Minorities in Science and Engineering: An Update.* NSF 92-303. Washington, D.C.: National Science Foundation, January 1992.

The Highway to Personal Professional Development

L. Carey Bolster

SOME experiences, for some reason or other, become etched in your mind, never to be forgotten ...

> It was my first day of school as a teacher. After three periods, I left the arena of teachers interacting with students to go to lunch.
>
> Lunch was a time I got to be with my colleagues. I immediately enjoyed them because we were a chance family, all of whom were brought together in a building called school. Teachers at the lunch table were unique in terms of their experiences, attitudes, and acceptance for change. In many ways we represented a cross section of our larger community, the community of professional educators.
>
> At lunch, while gobbling down our food and enjoying the precious few minutes of relaxation, the conversation always centered on school. A faculty meeting beginning at 3:30 on motivating students was to be held this Monday. The daily bulletin announced a fifteen-session in-service course every Wednesday, from 4:30 to 6:30. A speaker was coming to the teacher center, and funds were available for a substitute if we planned the lesson and found the sub. We heard comments such as "I'd like to take the in-service but my daughter has ballet that night and I have to drive her," "It's a lot of work correcting all those papers the sub gave. I'm not sure it's worth it," and "I'm not too motivated about this faculty meeting. I'm just tired."

I noticed that teachers did not buy into staff development because—
1. it was imposed on them;
2. it was an activity that was added onto their work schedule;

Carey Bolster is director of the Middle School Math Project, a service of PBS MATHLINE. He has been a teacher, department chair, supervisor, and coordinator of mathematics in the Baltimore County School System. He has had considerable experience in staff development and in curriculum development, implementation, and assessment.

3. it interfered with their personal life;
4. quite frankly, they felt it did not help them become better teachers!

I realized that you can't drive forward and constantly look in the rearview mirror! Part of the difficulty with staff development was its factory-like approach. This approach led to such well-meaning misconceptions as (1) all teachers need the same thing, so paint them in the same assembly-line fashion and (2) participating in more and more courses or hearing an expert telling what needs to be done will make teachers better equipped in the classroom. Consequently, professional development was considered an imposition—jump the hurdle, get your rewards, and forget it!

THE ON-RAMP

> *Late that afternoon of my first day, Dick Thomas, the science teacher, stopped by my classroom to see if I had survived. We sat in student chairs and shared "war stories." Since he was teaching periods of time, such as the Jurassic Period, he asked that I discuss large numbers and exponents in math class. "Of course," I said, wondering what the Jurassic Period was. "Be glad to do it. Could we meet tomorrow and talk about it?" This led to a yearlong series of "catch as catch can" meetings between Dick and me where we discussed curriculum, kids, parents, successes and failures, and a whole bunch of things about each other.*

Little did I know I had just entered the on-ramp to a lifelong trek on the personal professional development highway!

(From now on, I will often refer to personal professional development as *PPD*. Be sure to accentuate each letter with meaning! *PPD* Personal ... Professional ... Development. Say it pi times. I know that's irrational, but it is the true essence of teaching and learning.)

PPD NEEDS A JUMP START

Professional development is a dynamic, energy-producing, interactive experience in which participants examine and explore the complex components of teaching. The very heart of teaching demands that professional development be a continuous process of learning in which new knowledge is generated and embraced as the fine skills of teaching are employed, reflected upon, modified, and implemented. Effective professional development is based upon the belief that well-orchestrated programs and activities will result in improved instruction. Fundamental to this belief is that everyone involved in educating students benefit from professional development.

In my opinion, professional development has been given a low priority, while in fact, it is the keystone to the reform movement in mathematics education. If we are to achieve the recommendations of the trilogy of NCTM *Standards* documents, and other standards such as Goals 2000, we must jumpstart reform in the area of professional development.

We need to realize that personal professional development—

- is essential;
- takes time;
- affects the quality of instruction;
- determines implementation of systemic change;
- demands collaboration as teacher expertise is tapped;
- is so important that it needs to be on school time;
- needs positive support from policymakers, the public, and the profession itself.

Effective professional development initiatives—

- involve teachers, school and district supervisors and administrators at all levels, policymakers, and parents to ensure that necessary resources and support systems are available;
- empower teachers both to be responsible for their own vision of staff development and to recognize their potential impact as professional change agents;
- embrace collaboration and free exchange of ideas in which everyone's comments are important;
- recognize that change is difficult and needs to be nurtured in an environment where risk taking is encouraged;
- require knowledgeable and effective leadership at the state, district, and school levels.

IS MY ENGINE GOOD ENOUGH?

We are facing mandates from a variety of sources to upgrade the instructional program to meet standards, effectively implement algebra for all, use cooperative learning, implement writing across the curriculum, do interdisciplinary teaching, develop core curricula, use thematic approaches, effectively implement the inclusion model, and extend the knowledge base of every student so there is academic excellence for all—to mention just a few.

And to mention a few more of our mandates, we are expected to do excessive clerical work, make daily plans, hold parent conferences, and initiate alternative assessment. Is there life after school? Yes, there is life after school! With all this to do, I asked myself, "Does my engine have enough power?"

Well, you and I won't have the power unless we feel the need to change and improve.

After each class I taught, I asked myself, "Would I like to be a student in my class?" My answer was, "Not yet." This was an important question for me. I had a lot to learn ... still do ... not for someone else ... just me!

MERGING ONTO THE PPD HIGHWAY

I learned that when it comes to PPD, you can't just apply it like an exterior coat. It has to soak through and be polished, and the whole process must be repeated. In this way you will have a professional gloss!

The need for professional development comes from within. The desire to continue to develop professionally is a key component of being an effective teacher. Oh, by the way, the desire never stops! It accelerates through meaningful, consistent interaction and reflection. It is developing competence as a professional. It is a passion!

You are in control of the educational vehicle being driven in your classroom. You are the one! Professional development, like so many aspects of teaching, is a very personal thing. Regardless of all the regulations imposed by the system and others, the decision for professional improvement is solely up to you.

DIMMING THE LIGHTS

Remember the comment "Teaching is what happens when the teacher closes the door"? To a degree it is true. However, there are tremendous pressures, both internal and external, that teachers face when they close that door. One can no longer point to teachers and state that they are the only ones who have to change. The leaders within the organization of the school, district procedures and priorities, agendas from local and governmental agencies, and parental concerns influence what the teacher feels can be done and, in fact, can do. This multilayered complex environment often gets in the way of teachers' implementing changes that will significantly affect student learning.

I once heard a teacher say, "I would rather teach than learn." I felt sad inside because that teacher's light had been snuffed out by something ... or someone. We need not dim our professional development lights; we need to put on our brights. We can revitalize teachers' views by realizing that professional development is the act of building a culture of learning. We need to modify the workplace so that teachers, administrators, resource personnel, and the entire staff have a stake in the reform movement. Then we need to get the support of policymakers, parents, and the public.

DIFFERENT SPEEDS

When you enter the PPD highway, some drivers are moving rapidly, others are entering cautiously and progress to an accelerated speed, and some are just chugging along. Once on the highway, we notice that many want to pass and thus choose to change lanes.

It is obvious that any successful PPD initiative must provide the teacher with a wide variety of choices and options. The goal is to have something for everyone so that everyone is moving forward, even if at different rates.

However, the slower vehicles need to be ticketed. We must have a minimum speed when it comes to professional development. Too many students' futures are at stake. Remember, there is no maximum speed limit on the road to professional development, but there should be a minimum!

THE MULTILANE HIGHWAY

The hallmark of a quality comprehensive staff development activity is *breadth, depth, and follow-through*. If every teacher is to participate in professional development activities that directly affect the quality of instruction, we need a flexible game plan to ensure that each teacher is involved at some level. Through involving, nurturing, and giving recognition, significant professional growth experiences can occur.

The goal is to empower educators to be responsible for their own professional development by participating and growing in terms of pedagogy, content, and decisions that affect the learning culture.

CARPOOLING ON THE PPD HIGHWAY

As teachers we want to collaborate, we want to share ideas, we want to become better professionals, we want opportunities to take risks, we want our kids to succeed in mathematics and develop math power, and we will bend over backward to make it happen. We even spend our own money to buy materials for our students! Give each of us the chance to participate with others in the change process through a variety of professional development experiences. We can and will deliver!

Let's form professional car pools where teachers share the load, interact, and bond with one another. Let's make sure that this professional car pool has teachers, administrators, and resource personnel such as supervisors in it. Let's make sure that all know where they are heading, and let's give them the professional respect to make their own decisions as to the route they will take. Let's be sure we give them support when they have a flat! Let's form a series of car pools, a caravan with the goal of developing a teaching force that will enable students to reach higher standards.

If lasting benefits are to be derived, PPD experiences must be carefully orchestrated. This approach empowers teachers to actively participate in setting goals, to be a knowledgeable resource for others, and to assume a leadership role in the change process.

A FLEET OF MODELS

On the personal professional development highway there are many models, some of which enhance systemic reform more than others. There are many effective models of professional development. Only a few models are described below. Regardless of the models you are using, the point is that each person needs to experience a variety of models.

Site-Based/In-school PPD

There are few who would question the value of a school staff actively involved in determining the goals, the learning climate, and the curriculum designed to ensure student achievement and success. If we are to reach the students with the goals of systemic reform, it has to happen in the school. Teachers and administrators are the ones that have direct contact with students, and thus they should have greater control over their situation.

This model has these features:

- National, state, and school systemic initiatives adapted to meet the needs of the students in that school
- A faculty jointly agreeing on numerous issues relating to instruction and management
- Opportunity for growth as teachers assume more responsibility
- Opportunities for grade-level or subject teams to share ideas on a regular basis, to discuss curriculum and student progress, and to coordinate routine expectations

Tips: This PPD experience is more effective when—

- professional activities are designed to involve and prepare the entire staff for the more varied and complex roles they will play in restructured schools;
- roles are defined in a way that promotes and develops leadership among the entire staff in a professionally stimulating and open environment;

- sufficient time is provided, on a regular basis during the school day, for teachers to meet to fully implement the decision-making process at the school and grade level;
- resource personnel collaborate with the staff, inject new ideas, and provide a broad perspective.

The One-Day In-service Program

This model of staff development allows a teacher to get "off the highway" and take a side trip. When on such a short jaunt, one often sees an issue from a fresh view, and that can be stimulating!

This model has the following features:

- An opportunity to hear a presentation on a specific topic or issue by a panel or a noted authority in the field. The presentation provides information and stimulates discussion.
- A variety of activities or breakout sessions. Using some of the many reporting methods, participants share ideas.
- A platform for educators and others to interact with people outside their school

Tips: This PPD experience is more effective when—

- several interested teachers from a given school are involved;
- teachers anticipate what they hope to attain, reflect on the session, and make recommendations to their entire staff. The experience is not viewed as an assignment or a task but as an opportunity for professional leadership.

Conventions

Attending a convention can have many benefits to a school and to a district when groups of teachers have the opportunity to interact with others outside their school, district, and state.

This model features the following:

- An opportunity for extended networking and mutual support
- Insights into a wide variety of new techniques, content, and national initiatives
- Increased knowledge of the NCTM Standards—their meaning and implications for teaching and learning

Tips: This PPD experience is more effective when—

- participants, in a group, develop major areas of concern or interest (The program is examined, and sessions relating to these areas are identified. The group of teachers who attend sessions on the same major area then develop an action plan for sharing the ideas with others.);

- attendees come from a district so the system can benefit as well as the school;
- teachers serve as a nucleus for further district and school professional development activities.

The Cadre/Teacher-to-Teacher PPD

Building a cadre of teachers who are prepared to train other teachers is a model that is often used on the district level.

This model has the following features:

- A group of teachers from a variety of schools who collaboratively build and deliver professional development programs for other teachers
- Programs designed to further national, state, and district goals; develop expertise in certain areas such as technology; enhance understanding of content; and develop pedagogy
- Cadre teachers assisting local schools, teaching in-service programs, and providing a tremendous professional resource
- Cadre teachers forming close professional and personal bonds

Tips: This PPD experience is more effective when—

- leaders at the district and school levels collaborate with others to determine specific goals, procure funding, and select cadre members;
- cadre members reflect a cross section of the teachers in the district;
- the cadre teachers receive training utilizing good teaching techniques that can be used with other teachers;
- cadre members determine the breadth and scope of the professional development activities.

The Seminar

Seminars on a variety of topics are being used in many districts. Seminars provide opportunities for teachers pre-K–12 to investigate topics of interest.

This model features the following:

- A schedule of topics. Teachers choose to attend only those specific sessions that interest them.
- Teachers pre-K–12 participating in activities side by side, learning from each other, and gaining an appreciation and understanding of one another
- Flexibility for teachers who, because of commitments, may not be able to participate in a semester program

- A nonthreatening environment for teachers who have not been in the mainstream

Tips: This PPD experience is more effective when—

- the seminars are held throughout the year, cover a wide variety of topics, and incorporate active involvement;
- an informal count of the number attending each seminar can be taken in advance so arrangements such as the room size and sets of materials needed can be made to accommodate the size of the group;
- teachers receive credit for attending a given number of seminars (However, teachers can be given an extended period of time, say two years, to accumulate the number of sessions required. Records can be kept for each teacher using a spreadsheet.);
- opportunities are available for participants to share the knowledge gained with others.

PPD on the Information Highway

Emerging technologies are playing an ever increasing role in professional development as opportunities for telecommunication are becoming more available to educators. What once was considered a dream is a reality today, and the potential is staggering. Many teachers have a fear of technology, but the "point and click" computer format is so easy to use, and the benefits so great, teachers will want to participate.

This model has these features:

- Teachers' being able to participate in PPD activities any time or place convenient to them. This frees teachers to participate in long-range activities without having to be at a certain place at a certain time.
- Opportunities to make visitations electronically to other classrooms. Through a variety of technologies, teachers can observe other teachers incorporating aspects of standard-based instruction and can "revisit" the class as many times as desired.
- On-line interaction used with other educators in electronic learning communities. Computers equipped with modems, some of which are wireless, allow teachers to network within a region or across the country.
- Discussion of crucial educational issues through national plenary sessions, reports on new curriculum initiatives, and summaries of current information affecting education
- Delivery of "just in time curricula," in which individual requests are answered, using worldwide webs in combination with web crawlers, which seek information electronically. Teachers will be able to immediately obtain sample lesson plans, resources, and information.

Tips: This PPD experience is more effective when—
* educators are encouraged to participate and there is a climate in which professional educators and others embrace the idea that new opportunities require new solutions that do not necessarily fit the old mold;
* leaders in education establish partnerships, providing support for teachers by ensuring they have access to technologies such as a computer with a modem.

THE REST STOP

This is an exciting era in education. As professional educators we are committed to ensuring that each student has mathematical competence, has confidence in his or her ability to do mathematics, and is successful.

This means that more than ever, we, the professional community, need to be immersed in significant professional development. It is the very fiber of being an educator.

Enough of a rest, get back on the road!

. . . .

BIBLIOGRAPHY

Avila, Linda. "Teacher Empowerment through Professional Growth." In *Supervision and Site-Based Decision-Making: Roles, Relationships, Responsibilities, and Realities,* 1992 Yearbook of the Texas Association for Supervision and Curriculum Development, edited by Genevieve Brown and Jimmy Merchant, pp. 11–13. Houston: ASTD, 1992.

Collis, Betty. "Triple Innovation in the Netherlands." *Computing Teacher* (October 1994): 23–26.

Hirsh, Stephanie, and Gerald Ponder. "New Plots, New Heroes in Staff Development." *Educational Leadership* (November 1991): 43–48.

National Council of Supervisors of Mathematics. *Supporting Leaders in Mathematics Education: A Sourcebook of Essential Information.* Golden, Colo.: The Council, 1994.

National Council of Teachers of Mathematics. *Professional Standards for Teaching Mathematics.* Reston, Va.: The Council, 1991.

National Education Commission on Time and Learning. *Prisoners of Time.* Washington, D.C.: The Commission, 1994.

National Governors' Association. *Professional Development for Educators a Priority for Reaching Standards.* Washington D.C.: The Association, 1993.

National Staff Development Council. *National Staff Development Council's Standards for Staff Development: Middle Level Edition.* Oxford, Ohio: The Council, 1994.

The National Board for Professional Teaching Standards: Making Professional Development "Professional"

James A. Kelly

F OR far longer than anyone in the education community may care to acknowledge, the issue of professional development for teachers has been a sore point. One might even venture to characterize it as an embarrassment. It is true that a majority of the nation's school systems have traditionally paid lip service to the notion of professional development, requiring some form of periodic in-service activity of their teachers. But the efforts lacked a substantive vision of accomplished teaching to shape and guide professional development activity. What's more, rarely was either the structure or the content of these offerings chosen or designed by the teachers themselves. Little wonder, then, that many excellent, experienced teachers with years of in-service training behind them remain skeptical of current efforts at professional development. Their own experiences with "professional development" have been neither "professional" nor "developmental."

Instead, these experiences have not only been discouraging to individual teachers, they have also served to raise serious and disheartening questions about America's attitude toward the so-called "profession" of teaching. On the one hand, we entrust teachers with our greatest resource, our children, and expect them to perform a series of daily classroom miracles. On the other

James Kelly is president of the National Board for Professional Teaching Standards. He was a program officer at the Ford Foundation, where he directed programs in education finance reform, and has served on the faculty of Teachers College, Columbia University. He began his career in education as a teacher and school administrator.

hand, we pay them poorly, treat them with little or no respect, deny them adequate materials and current technology, and once they are in the classroom, expect them to fend for themselves in a class full of children who might have experienced abuse and neglect, poverty, pregnancy, alcohol and drug addiction, violence, gangs, and persistent racism. And these are the children who must be taught to succeed in an increasingly diverse society, in a rapidly changing technological world, and in a competitive global economy.

Recently I was chair of a panel sponsored by the Charlotte-Mecklenberg (N.C.) Schools. Our goals were to identify the elements of a "world class" school system and to establish a plan of action leading to genuine school reform. Our published recommendations contain this description of a teacher's day (Murphy 1992, pp. 52–53):

> Punch a clock. Sign in and out of the building. Thirty minutes for lunch. Schedule bathroom breaks. No access to the school building unless the students are there. No private offices. No phones for private calls. No time to confer with colleagues. Handle bus duty, corridor duty, cafeteria duty and playground duty. Imagine asking lawyers, doctors, or college professors, for that matter, to put up with such conditions for one week, let alone a lifetime.

The truth is that although we *call* teaching a profession, and we make daunting demands on teachers, in important ways we treat teachers like blue-collar workers rather than professionals. In fact, the private sector invests heavily in the continuing development of the skills of its front-line workers in order to strengthen productivity, quality, and competitiveness, but the public sector fails to make a similar commitment to its teachers.

When money is scarce, one of the first areas to be cut is professional development. In-service training for teachers is considered expendable. When professional development programs are in place, most lack comprehensive standards and substance. According to Berkeley's Judith Warren Little (Little 1993, p. 22):

> Much "staff development" or "in-service" communicates a relatively impoverished view of teachers, teaching, and teacher development. Compared to the complexity, subtlety, and uncertainties of the classroom, professional development is often a remarkably low-intensity enterprise. It requires little in the way of intellectual struggle or emotional engagement, and takes only superficial account of teachers' histories or circumstances.

In too many schools, what passes for teacher development programs are infrequent afternoon or after-school sessions, pulled together at the last moment and offered when teachers are exhausted. Furthermore, the very structure of schools places obstacles in the way of professional collegiality. The organization of assignments and the physical layout of the buildings promote isolation, not professional conversation. And the typical salary schedule rewards simple longevity and encourages the accumulation of graduate credits without regard to the relevance, quality, or rigor of the course.

It is clear that the education community needs to rethink professional development and to send a signal to teachers that professional development does not end the day they begin teaching. If America is genuinely committed to making substantive, permanent changes in the nation's schools, the initiatives must derive from a substantive understanding of teaching and must involve teachers, for teachers are at the heart of education. As the nation embarks on a process of redefining the goals and purposes of education and establishes high and rigorous standards for what students should know and be able to do, we must work toward creating a teaching profession equal to that task—a profession of lifetime learners.

To create such a profession, we must design professional development initiatives with the following qualities:

1. They must be teacher driven, rather than imposed from above.
2. They must be clearly focused both on standards for excellence in teaching and on the same vision of standards that we set for our students.
3. They must emphasize student learning and presuppose high expectations for all students.
4. They must be a priority for both teachers and administrators. As such, they must be sustained, rather than one-time or weekend events, and continuous over the course of a teacher's career.
5. They must have a strong clinical component. That is, teachers should be observing, and be observed by, highly accomplished teachers; practicing new techniques; incorporating new content; using new technological tools; and discussing their performance with exemplary teachers and their peers.
6. They must encourage teachers to analyze, reflect on, and learn how to improve their actual teaching practice.

THE VISION OF THE NATIONAL BOARD FOR PROFESSIONAL TEACHING STANDARDS

The same critical issues that have been raised thus far in this essay were the subject of *A Nation Prepared: Teachers for the 21st Century*, published in 1986 by the Carnegie Forum on Education and the Economy's Task Force on Teaching as a Profession. The task force report concluded that to facilitate a rise in educational performance and achievement and the creation of a teaching profession equal to the challenge, America must make sweeping changes in education policy. Foremost among the Carnegie Task Force recommendations was the establishment of a National Board for Professional Teaching Standards (NBPTS) to establish high and rigorous standards for what accomplished teachers should know and be able to do, to develop and operate a national voluntary system to assess and certify teachers who meet these standards (National Board Certification), and to advance related education reforms for the purpose of improving student learning in American schools. The National Board was established in 1987. It is governed by a board of sixty-three members, the majority of whom are elementary and secondary school teachers. Several prominent mathematics teachers sit on the board, and two of the board's founding members were recent past presidents of the National Council of Teachers of Mathematics (NCTM).

The vision of the National Board is shaped by an image of teaching that is complex but is grounded in knowledge that can be learned and applied by most teachers. Teachers are informed and principled decision makers, professionals who make myriad decisions influenced by a host of complicating factors. Like the anthropologist in the field or the scientist in the laboratory, the teacher in the classroom is guided by working hypotheses, educated guesses about the nature of what happens. These working hypotheses are constantly tested, revised, and revamped in light of what is known about teaching and by teachers' own knowledge and experience.

Take, for example, a teacher working with students on how to use decimal concepts and operations to model and solve problems. Even before teaching the lesson, the teacher has had to make many decisions, such as which concepts are most important in light of the new technologies and societal priorities, or how each concept fits into the overall sequence of instruction. At any point during the lesson, the teacher is weighing and acting upon a shifting complex of information:

- What does the student already know?
- How can new knowledge be linked to concepts and ideas the student already knows?
- How can the student accommodate or compensate for areas of weakness while moving ahead to learn new concepts and strategies?
- What are the special strengths and weaknesses of the learner in abstract and linguistic reasoning?
- How well does he or she understand numbers and the operations performed upon them?
- What interests does he or she have that might be used in developing interesting problems? (For example, could enthusiasm for baseball be used to transform a concept into a question of batting averages?)
- How is that one learner situated in the classroom or in the smaller group within that classroom?
- What tasks can be used so that students will gain an understanding of the most important concepts?

- How can manipulative materials or technological tools like calculators and computer software be incorporated to support student learning of this concept?
- How will the knowledge be assessed?
- What time of day is it?
- How long will the student's attention span last?
- What does the teacher do if the planned approach for the concept simply does not work?

And, of course, each working hypothesis or decision is never made for only one learner at a time. The teacher makes decisions for twenty-five or more students simultaneously, weighing the best answers for the group. Amidst this Babel of competing interests, the teacher is charged with selecting plausible solutions.

Such decision-making skills are not the hallmark of every occupation; rather, they are characteristics of a *profession*—where practitioners make judgments by drawing on professional skills and knowledge learned from both training and experience. The more one has the chance to hone these decision-making skills, the better they become. Thus professional experience becomes a critical component in developing and evaluating professional competence.

In light of this vision, since its inception the National Board has operated under the assumption that accomplished teaching is the result of a continuum of learning and professional reflection. Consequently, the NBPTS has created a system of National Board Certification in which each facet of its two critical elements, standards and assessments, has been designed to reflect this assumption. A teacher who has taken part in National Board Certification activities has, in essence, experienced a series of valuable professional development activities that deepen knowledge and promote reflection.

NATIONAL BOARD CERTIFICATION: HOW IT WORKS

The National Board Certification system is an important step in the direction of strengthening America's education system. As we have seen, although America pays lip service to the notion of teaching as a profession, unlike medicine, architecture, or accounting, until now the teaching profession has not codified the knowledge, skills, and dispositions that account for professional excellence. The failure to do so has left room for continuing misconceptions about what constitutes accomplished teaching. Thus, far too many Americans—even administrators and teachers themselves—still believe that any modestly educated person with a nurturing instinct has the requisite qualifications to teach. Meanwhile, some educators focus on knowing one's subject, others argue for child-centered teaching, still others for a methods approach. The truth is that all these dimensions have validity, although even in combination they fail to do justice to the breadth of a teacher's responsibilities. Other necessary skills for an accomplished teacher are accurately evaluating student needs and progress, translating complex material into language students understand, adapting curriculum and strategies to reflect the influence of current technologies, selecting worthwhile tasks that promote understanding, exercising sound and principled professional judgment in the face of uncertainty, and acting effectively on such judgments. The experienced, accomplished teacher also takes on the dual role of education ambassador and role model to the community by making herself or himself available outside the classroom to parents and professional colleagues.

Until now, most attempts to recognize accomplished teachers not only have been characterized by limited teacher involvement in their origin, but also have not been based on high standards of professional practice. These efforts, as well as traditional "teacher of the year" awards, often decided from outside the profession, are greeted with considerable skepticism by teachers themselves. The absence of a credible and accepted method of recognizing outstanding teaching reinforces the notion that anyone with a fondness for children can teach—that the profession does not take itself or its responsibilities seriously. These factors help to shape an image of teaching that is often frustrating to those who elect to venture into it and particularly discouraging to those with precisely the qualities the nation needs to strengthen its schools.

A system of National Board Certification that commands the respect of the profession and the public can make a critical difference in how communities view their teachers, in how teachers view themselves, and most important for the purposes of this essay, in how teachers can improve their teaching practice continuously throughout their careers.

National Board Certification differs from state licensing, sometimes called "state certification," in four critical ways. First, a state license is mandatory for teaching within its jurisdiction and indicates that the licensee satisfies entry requirements. In contrast, National Board Certification will be voluntary. Second, licensure sets minimum entry standards, whereas the National Board will recognize accomplished teaching based on rigorous professional criteria. Third, state licensure requirements vary from state to state, whereas National Board Certification standards are uniform across the country. Finally, and perhaps most important, National Board Certification is developed *by* teachers *for* teachers, not established by those outside the profession. It will be awarded to those who pass a series of performance-based assessments, involving activities both at the candidates' schools and at designated assessment centers.

National Board standards and assessments are based on a fundamental philosophical and policy framework expressed in the following *five core propo-*

sitions establishing what accomplished teachers must know and be able to do and what accomplished teaching practice is:

- Teachers are committed to students and their learning.
- Teachers know the subjects they teach and how to teach those subjects to students.
- Teachers are responsible for managing and monitoring student learning.
- Teachers think systematically about their practice and learn from experience.
- Teachers are members of learning communities.

To accommodate the diverse nature of the American teaching experience and the broad range of expertise and specialization expected of teachers, NBPTS is establishing standards in more than thirty certification fields. The certification fields recognize developmental level of students, subjects taught, and context of teaching, and collectively will cover all K–12 teachers, including those who teach students with special needs. NBPTS standards committees are composed of highly experienced, outstanding professional educators; practicing classroom teachers make up the majority of and chair all the standards committees. They are engaged in establishing high standards that address both the unique characteristics of each certification field and the overall National Board five-point statement.

In keeping with its vision of teaching as a highly skilled, collegial enterprise involving complex decision making, NBPTS has designed an assessment system that is performance-based and employs a broad range of innovative assessment strategies. At the same time, it will be professionally credible, publicly acceptable, legally defensible, administratively feasible, and economically affordable. The initial model for the assessment involves a series of exercises organized in two modules, one requiring a portfolio collected in a candidate's school setting and the second requiring data collected later at an assessment center. The school-site portfolio consists of reflective documentation of teaching and student learning and videotapes of classroom instruction. The assessment-center module focuses on structured interviews, simulations, written work, and assessments of knowledge of subject matter and content-specific pedagogy.

During the 1993–94 school year, over 500 teachers took part in a national field test of the first two certificates under development: Early Adolescence/ English Language Arts and Early Adolescence/Generalist. A field test network (FTN) was established to test the assessments, recruit initial candidates, and develop a variety of professional development models. The FTN was composed of 112 school districts and colleges and provides access to a nationally representative sample of teachers covering approximately 7 percent of the nation's elementary, middle, and secondary school teachers.

After submitting their portfolios, the field test candidates were encouraged to evaluate their experiences. Their responses indicate they believe that the process of National Board Certification is a rigorous professional development activity with concrete, positive effects on teaching practice.

One Early Adolescence/Generalist candidate was especially enthusiastic about the videotaping component of the assessment, seeing it as a "fantastic vehicle for reflection." In assembling his portfolio he called on colleagues to help in the taping and to review the tapes. "These discussions have been invaluable to me in evaluating the lessons. They have also benefitted my colleagues, giving them a new insight into a precise and yet faithful method of self-evaluation."

An English language arts candidate reported, "The process was rigorous and demanded careful analysis and reflection about our teaching practices. We have learned so much. We applaud the efforts of the National Board to design an assessment process which reflects the complexity of teaching and allows us to be judged by our peers."

Still another Generalist candidate explained, "The addition of 'reflective practice' to my daily practice has turned my otherwise good teaching into extraordinary teaching; analyzing setbacks has now become critical for continual self-improvement." The insights she gained from assembling portfolio materials on three students have led her to adapt this technique across the board. "I've scheduled portfolio conferences with all my students, and I am redesigning my day to include more portfolio interaction."

The full system of National Board Certification will be phased in over the next five years. Development and field testing for the Adolescence through Young Adulthood (ages 14–18+)/Mathematics certificate will occur over the next two years. Teachers will be able to participate in the National Board Certification process for this certificate during the 1995–96 school year. Development of the Early Adolescence (ages 11–15)/Mathematics and Middle Childhood (ages 7–12)/Mathematics certificates will follow. Teachers of students at these developmental levels can anticipate participation in the process in 1997–98.

NATIONAL BOARD CERTIFICATION AS A REFLECTION OF NCTM's *PROFESSIONAL STANDARDS FOR TEACHING MATHEMATICS*

As well as offering a unifying view of accomplished teaching across the disciplines, the National Board for Professional Teaching Standards is committed to designing a certification procedure that acknowledges the importance of each of the disciplines, including, of course, mathematics. The National Board has called upon recognized experts from the field of mathematics teaching to help shape subject-specific standards to be used for certifying accomplished mathematics teachers who teach children at these different developmental levels: middle childhood (ages 7–12), early adolescence (ages 11–15), and adolescence and young adulthood (ages 15–18+). These experts include excellent mathematics teachers, other mathematicians, and mathematics teacher educators.

The picture of accomplished mathematics teaching that emerges from the NBPTS mathematics standards is wholly consistent with that described in the National Council of Teachers of Mathematics' (NCTM) *Professional Standards for Teaching Mathematics* (1991). The NBPTS mathematics standards strongly emphasize the importance of teaching challenging mathematics to all students, including those who have not been successful in mathematics or in school. This focus on access to mathematics for all students is a cornerstone of accomplished mathematics teaching and is recognized as such by both the National Board and NCTM. The National Board also recognizes that all too often, mathematics has been used as a way to sort students out of such opportunities. The National Board's mathematics standards leave no doubt that a critical role for mathematics teachers is to find ways to allow previously unsuccessful students to wrestle with, and learn from, challenging mathematical tasks. Working with all students, which implies working with all students individually and in small groups, is viewed as a challenging, but necessary, component of professional mathematics teaching today.

Those standards also recognize the importance of selecting meaningful and appropriate mathematical tasks and activities for students, providing access to computers and calculators as problem-solving tools for students, and managing how students interact with those tools, with those tasks, and with each other. Although it is clear that an accomplished mathematics teacher must play many roles and use many strategies, the emerging picture of a mathematics teacher is moving toward that of a facilitating coach and mentor and away from that of a well-prepared lecturer. Furthermore, he or she is comfortable with the latest technological advancements in the field. This shift is reflected in the NBPTS mathematics standards themselves and will be a component of the assessment program to measure those standards. For example, as part of the school-site portfolio of the current Early Adolescence/Generalist assessment package, candidates are asked to become familiar with the SimCity computer simulation and evaluate it as an instructional resource. As an assessment-center exercise for mathematics candidates, teachers might be asked to discuss how technology has affected the curriculum and teaching of mathematics and describe how they might have students use a graphing calculator to explore the concept of "line of best fit." Or, they might be asked to read and discuss a journal article on the use of a particular computer software program to teach concepts related to quadrilaterals. The National Board is exploring the possibility of having mathematics candidates bring graphing calculators to the assessment center and providing word processors to candidates to use in responding to assessment center questions.

A teacher who has remained at the level and style of mathematics teaching that might have been effective twenty years ago may have to modify and improve his or her practice to be able to demonstrate proficiency across the breadth of skills, strategies, and understanding for working with all students reflected in the NBPTS mathematics standards.

The NBPTS mathematics standards for different student developmental levels, including specific standards on knowledge of mathematics, knowledge of students, and knowledge of mathematics pedagogy, paint a picture of professional development that can serve as a road map for teachers planning collegial projects, courses, readings, institutes, or nonroutine professional alternatives they might choose for their own professional growth. This substantive road map can also serve to guide those who provide or create professional development opportunities for mathematics teachers, whether at the preservice or the in-service level.

Equally important, the National Board's mathematics standards emphasize reflection and analysis that are not only expected in daily practice but are experienced through the process of preparing for and participating in the NBPTS assessment itself. Both the National Board and the NCTM have identified this kind of professional reflection as a critical component of professional mathematics teaching. Choosing how to participate in National Board mathematics certification and preparing for that participation can provide an excellent opportunity for reflection and analysis about one's practice, according to teachers who field-tested the first two certificates.

NCTM's *Professional Standards for Teaching Mathematics* (1991) emphasizes the active role teachers should take in their own professional development, from their choices about the kinds of activities in which they engage to their interactions with colleagues. When an accomplished mathematics teacher chooses to become a candidate for National Board Certification, he or she is demonstrating the kind of professionalism described by NCTM.

Furthermore, NCTM's standards describe the role that should be played by policymakers and leaders both from the profession itself and from outside the profession to ensure that mathematics teachers are engaged in the kinds of rigorous professional development activities that are commensurate with the needs of a strengthened teaching force.

A CLOSER LOOK AT NBPTS MATHEMATICS ASSESSMENTS

Whether in one of the mathematics-specific fields or within the broader context of the mathematics included in one of the generalist fields, the emerging assessments are proving to address powerful aspects of mathematics instruction. Early versions of field-test assessments in mathematics clearly demonstrate the potential for this kind of assessment to look at mathematics teaching far beyond the assessment of individual standards such as Knowledge of Students or Knowledge of Mathematics. Within the exercises, a teacher has the opportunity to present actual samples of students' work, reflecting on what the teacher learned about each student from student responses to a particular assignment. The teacher follows students' progress over an extended period of time, describing the focus of the mathematics instruction and discussing how instruction was modified on the basis of what the teachers learned from students' work. Teachers demonstrate an in-depth understanding of mathematical concepts addressed in the instruction, while also showing a knowledge of how their students learn. In order to do well on the exercises, teachers need a broad and deep understanding of mathematics, so that they can demonstrate that they have chosen among several sound conceptual approaches on the basis of the topic and of students' needs and interests. They are called on to relate the mathematics they are teaching to other domains within the discipline, making connections that help students develop a sense of the "big ideas" of mathematics. Candidates must possess a strong professional background in mathematics, not so much in terms of specific coursework but rather in terms of their comprehensive understanding of the most significant themes and ideas of mathematics.

In other exercises currently being field-tested, teachers describe the in-depth development of a unit of instruction. They demonstrate how well they understand important mathematical ideas, how they would use tools like calculators or manipulative materials, how they adjust instruction on the basis of student responses, and how they develop and use a variety of appropriate student assessment measures. Successful candidates use their knowledge of mathematics in concert with their knowledge of the instructional process and the broader concepts of how people learn mathematics. They model an understanding of the kind of mathematics described in NCTM's *Curriculum and Evaluation Standards for School Mathematics* (1989) and the kind of teaching described in NCTM's *Professional Standards for Teaching Mathematics* (1991).

NATIONAL BOARD CERTIFICATION AS A CATALYST FOR CHANGE

The National Board for Professional Teaching Standards and its National Board Certification system can be viewed as a dramatic means of professional development reform and can lead to a new kind of school as well as a new kind of teacher. The National Board Certification process itself is designed to encourage professional collegiality and leadership. It supplies teachers with opportunities for self-renewal, which is a major prerequisite for meaningful change.

The assessment requires that candidates demonstrate their ability to take risks, to solve problems, to collaborate with their colleagues. Most important, it requires serious professional reflection.

Preparing for National Board Certification requires engaging in the kinds of activities, both alone and in a collegial setting, that lead to stronger teaching. Teachers who have gone through the National Board Certification process have observed themselves in action, discussed their teaching with colleagues, reflected on the rationale behind their classroom practice, and evaluated the degree to which their students have acquired specific knowledge and skills.

To document the potential of the process for strengthening teaching practice, NBPTS recently engaged in a professional development project. A selected group of teachers used the National Board's standards as a lens through which to observe and judge their own teaching performances. They also attempted to identify the kinds of evidence that can be used as documentation to meet the standards and stand for National Board Certification. The project brought together a group of exceptionally accomplished teachers. A sampling of their letters of application to the project is evidence of their commitment and professionalism:

- "I have always believed we should be treated as professionals who bring skill and knowledge to our job. However, in many instances our profession is looked down upon. Perhaps having specific standardized criteria for certification will show that we are a concerned and introspective group performing to the best of our abilities ... the criteria should come from the people in the trenches on a day-to-day basis."
- "This grant is not only an opportunity for us to assess what we are doing, but to make a statement regarding what we feel is excellence in teaching and to make an impact on future members of our profession."

Collaborating with other candidates in preparing for the assessment encourages professional collegiality, while the feedback offered to all candidates should lead to stronger practice and increased insights regardless of the individual outcome or the degree to which individual teachers are interested in taking on new professional responsibilities in their schools.

CONCLUSION

The traditional means for acknowledging teaching excellence has been to move the teacher up and out of the classroom. With a system of National Board Certification in place, National Board Certified Teachers will be well qualified to assume interesting and challenging leadership roles without leaving the classroom. These teachers will be confident observing their colleagues and equally confident being observed; they can engage in substantive discussions about teaching; they can work with others in designing new courses and curricula; and they will understand how to undertake evaluation procedures and serve as mentors to colleagues.

Furthermore, a fair and rigorous National Board Certification system will offer school boards and administrators an objective means of identifying accomplished teachers and making full use of their knowledge and expertise. It will do so without running the risk of incurring charges of cronyism and without removing accomplished teachers from the classroom. Making use of a system that identifies accomplished professionals in a fair and trustworthy manner can help free the schools from a structural straitjacket and enhance the likelihood that all school resources—time, people, money—are marshaled more effectively on behalf of student learning.

Such a system can motivate more teachers to engage in substantive professional development activities and to work on improving their practice in order to advance. Most important, by demonstrating that teaching is a challenging professional endeavor which is treated with respect and seriousness by its own practitioners, by policymakers, and by the public at large, it can attack the twofold problem within America's education system: how to attract excellent teachers to the elementary and secondary schools and how to keep them in teaching once they are there.

I conclude this essay with a comment from one of those excellent teachers (and an Early Adolescence/English Language Arts field test candidate):

> With the test over ... not only did I feel an enormous amount of self-satisfaction, I also felt that the teaching profession was on the verge of a breakthrough. The questioning, and the process that I learned and worked with, have not left me. I work as a mentor and as a cooperating teacher, and I make sure that these young people just beginning in the profession have time to sit, reflect, retool, and try again. Undoubtedly there will be some colleague who will come to me in the future and ask if he or she should take this test to become National Board certified, and I will respond with a resounding yes.

· · · · · · · · · · · · · · ·

NOTE

These five propositions are explicated at length in the National Board's policy statement, "What Teachers Should Know and Be Able To Do." The document can be obtained by writing to NBPTS, 300 River Place, Suite 3600, Detroit, MI 48207.

· · · · · · · · · · · · · · ·

BIBLIOGRAPHY

Carnegie Forum on Education and the Economy. *A Nation Prepared: Teachers for the 21st Century: The Report of the Task Force on Teaching as a Profession.* New York: Carnegie Corp., 1986.

Little, Judith Warren. *Teachers' Professional Development in a Climate of Educational Reform.* NCREST Reprint Series. New York: Columbia University, 1993.

Murphy, John A. *The Charlotte Process: Reclaiming Our Legacy. City: Publisher,* 1992.

National Board for Professional Teaching Standards. *Toward High and Rigorous Standards for the Teaching Profession,* 3rd ed. Detroit: The Author, 1991.

National Council of Teachers of Mathematics. *Curriculum and Evaluation Standards for School Mathematics.* Reston, Va.: The Council, 1989.

____. *Professional Standards for Teaching Mathematics.* Reston, Va.: The Council, 1991.

Schwartz, Janet. Speech given to National Education Association Conference on Assessment, San Francisco, 29 June 1993.

South Brunswick National Board Project. Progress Report, 30 January 1993.

CONTENT . . .

THE first question faced in designing a school mathematics program is simple: What mathematical concepts, principles, techniques, and reasoning methods are most important for students to learn? Wise answers to that question will reflect a broad and deep understanding of mathematics and its many important applications, and that understanding must be based on a vision of the discipline that includes the projection of its future development. But a vision of mathematics as a discipline is only one aspect of planning and implementing curricular change—mathematics educators must see their subject in the full academic and social context of schooling. Furthermore, the design of a new curriculum is only the first step on a challenging road to genuine reform of the mathematics that young people experience in their schooling.

The essays in this section address fundamental issues in mathematics curriculum development by reflecting on recent and projected developments in the discipline, by thinking about the implications of those developments for school instruction, and by analyzing experiences to find effective strategies for curricular change. The first essay presents a broad vision of contemporary mathematics and its applications and then draws inferences about the curriculum for school mathematics by analyzing mathematics at work. The second essay looks at problems of mathematics curriculum reform from the perspective of a state education administrator, showing how mathematics fits into the broader picture of school programs and change. Then the third essay identifies crucial contemporary curricular issues in school mathematics and analyzes the processes by which forward-looking curricula can be developed and put into practice.

—James T. Fey

Mathematics for Work and Life

Lynn Arthur Steen & Susan L. Forman

Seventeen-year-old Ramona, like many of her friends, was bored with school—with books, with classes, and especially with mathematics. So she jumped at the opportunity to enter an apprenticeship program in which she would spend half of every week working in a local machine-tool shop. Here she would learn to use lathes and drill presses to make useful things rather than waste time on math problems with no earthly purpose that she could discern.

One month after beginning the apprenticeship program, Ramona had to set up a computer-controlled lathe to make a ring for a shaft that had been brought in for repair. First she checked the cutting tool to be sure that it was set for the required fifteen-degree angle. Then she identified the line in the computer code telling where the lathe would make its first cut on the circumference. She edited the code to include an offset of five mils (0.005"). She then executed the code and checked the gap with a five-mil shim. Because the fit was too loose, she corrected the offset and repeated the process. When the machine stopped within the required tolerance, she reset the x-coordinate of home by the difference between the thickness of the shim and the amount of the offset. She then repeated the process to reset the y-coordinate, made one last check with the five-mil shim, and started the lathe.

The next day, working with the automated drill press on a custom-designed fixture, she checked her setup by executing the computer code in single-step mode. The drill holes were positioned perfectly on the first side, but when

Lynn Steen is a professor of mathematics at Saint Olaf College in Northfield, Minnesota. He has served as president of the Mathematical Association of America, as chair of the Council of Scientific Society Presidents, and as executive director of the Mathematical Sciences Education Board.

Susan Forman is a professor of mathematics at Bronx Community College of the City University of New York. She has served as first vice-president of the Mathematical Association of America and as director for postsecondary programs of the Mathematical Sciences Education Board.

she flipped the fixture over, she discovered the holes were way off. She found the line in the code responsible for this step, studied the printout of the program near this line, and compared it with the earlier section of the program that had worked perfectly. She quickly discovered that the programmer had copied the *x*- and *y*-coordinates from the first part of the program, even though the parts would not be in the same position when the fixture was flipped over. So she edited the program at the computer console, ran the program in single-step mode to ensure that it would work as desired, and then executed a full run. Finally, she notified the programmer of the error she had discovered and requested that a revised program be installed in the drill.

Ramona's experience with today's technological workplace illustrates mathematics in context—in this case in the unlikely environment of a job shop. Others experience mathematics in equally unexpected situations, both at home and at work:

A homeowner is planning a built-in shelf unit for a remodeled kitchen. The current plan calls for 1″ × 8″ oak boards of four different lengths: four to six at 2′10″, three to five at 4′6″, seven or eight at 5′9″, and two or three at 6′6″. The local lumber yard sells oak boards in lengths of 6′, 8′, 10′, 12′, and 16′, but the shorter lengths are more costly per board foot. The homeowner needs to decide what combination of shelf lengths and purchased boards will yield the most economical shelf unit.

. . . .

A financial assistant in a major health maintenance organization (HMO) has been asked to make projections about changes in payments that might be expected if certain requested changes were made in one of the firm's corporate accounts. The company wants to increase copayments and broaden the scope of coverage, all without changing premiums. The assistant calls up a standard spreadsheet template that covers all the HMO group policies and locates the cells where projection calculations are made. He will need to study these cells to be sure he understands how their formulas now work, then modify them to reflect the proposed changes, and finally run several test cases to be sure that his changes accomplish what he intended.

. . . .

A truck driver for a local appliance company has to plan the day's schedule and load a van whose cargo space measures 4′3″ × 5′6″ × 7′. The day's deliveries include four refrigerators, each in a 34″ × 34″ × 68″ carton; three stoves, two in 32″ × 28″ × 50″ cartons and one in a 32″ × 28″ × 74″ carton; and a large freezer in an 80″ × 30″ × 30″ carton. The driver also has to pick up two television sets and a washing machine that have to be repaired. Before beginning, the driver will use a map of the delivery area to plan the day's deliveries in order to notify each customer of the estimated delivery times.

Mathematics in context is, typically, very concrete but not necessarily very straightforward. Problems arising from real situations generally can be solved in a variety of ways and do not necessarily have unique "correct" answers. Several strategies may be "good enough," even if one technically may be a bit better than others. Mathematics at work in ordinary contexts typically involves real data with realistic measurements expressed in common units. The technical skills required to deal with these data are relatively elementary—measurement, arithmetic, geometry, formulas, simple trigonometry. The problem-solving strategies, however, often require a cognitive sophistication that few students acquire from current school mathematics: planning and executing a multistep strategy; consideration of tolerances and variability; anticipation and estimation of relevant factors not immediately evident in the data; and careful checking to assure accuracy.

MATHEMATICS FOR WORK

Authentic problems from life and work, unlike the short "template" exercises found in textbooks or classrooms, reflect the way mathematics is used in the ordinary world—as a rich source of higher-order thinking based on lower-order mathematics. Thus, we begin with a few simple declarations:

The chief purpose of school mathematics is to prepare students for work. All students—including those who go to college—will work and will use mathematics in their work and throughout their lives. Moreover, the nature of work—especially in desirable jobs—is becoming ever more mathematical. Thus the major focus of school mathematics should be on those mathematical activities that people will use in their lives and work.

Mathematics in school should closely resemble mathematics in work. We know from years of unsuccessful experience that for most students, decontextualized learning does not last. Students retain what they learn from their own efforts to address challenging problems that arise from situations that resonate with their own interests. Workplace applications not only provide an apt context for school mathematics but also exercise sophisticated concrete skills in a manner that provides an effective, gradual approach to the abstract thinking necessary for advanced mathematics.

The mathematics required for work can provide strong preparation for college. The renewal movement in college mathematics, led by calculus reform, emphasizes many of the key principles that characterize mathematics in the workplace: interactive learning, real-world models, and multiple representations (numerical, graphical, symbolic, verbal). Moreover, solving authentic problems from mathematics in the workplace requires the successful execution of multistep strategies—just the kind of skill undergraduates need, especially for the "reformed" expectations of college mathematics.

Placing priority on mathematics in the context of work is meant not to supplant the traditional goals of school mathematics but to provide a more effective entrée to the mathematics that suits students' needs and interests. As students begin to see that mathematics is useful to them, they will become receptive to learning more-advanced mathematics. Mathematics learned through a robust curriculum that emphasizes skills useful in work can also provide appropriate levels of mathematical literacy for citizenship and lifelong learning. Thus, of the multiple goals often cited for school mathematics, preparation for work is the one with the greatest leverage to achieve all the others.

OLD SKILLS, NEW JOBS

Pressures for high-quality mathematics education take two forms: societal and personal. The newly competitive international economy supports the societal argument (Reich 1992) that national wealth in the future will depend on intelligence and education as much as on natural resources. On a personal level, the desire for high-quality jobs will, primarily because of the widespread impact of computers, require increasing levels of mathematical literacy. This need is most urgently felt by those whom Robert Reich calls the "anxious class," primarily residents of large cities and urban areas where there is a growing gap between wealth and poverty.

The employment opportunities growing most rapidly are those that require advanced technical skills. Decision making and quality control increasingly are becoming part of job expectations at all levels, from the production floor to management offices. As companies expand or contract, employees are often expected to become involved in corporate planning. And as workers move from job to job and as their career interests change, they often drop in and out of postsecondary education, particularly at two-year colleges. Students returning to education from the workplace will have different expectations, based on personal experience, of what they need to learn. This trend will most likely increase as people realize that their old skills are not well suited to new jobs.

At an earlier time, when people rarely changed jobs and occupational skills were well defined, school mathematics branched in the middle grades into two tracks—a short "terminal" track for students not planning postsecondary education and a full four-year curriculum for those called "college intending" in the NCTM *Curriculum and Evaluation Standards* (NCTM 1989). That time of simple, fixed career paths is long gone. NCTM now recommends that the first three years of high school mathematics be designed for *all* students, whether or not they plan to attend college.

Since historically the emphasis of high school mathematics was on preparing students for college, higher education was able to set the standard for the upper years of school mathematics—and it did so by making calculus the rite of passage in mathematics for all who aspired to professional careers. But now, as College Board President Donald Stewart argues (Stewart 1994), higher education's increased diversity makes it both less able and less likely to shape secondary education. Consequently, concerns for work-intending as well as for college-intending students—for an education for work and for life—take on greater prominence in determining priorities for school mathematics.

NEW MATHEMATICS, OLD SKILLS

Our assignment in this paper was to think boldly about "new directions" for the content of school mathematics. One can look for guidance at deep ideas from within mathematics, at new ideas from today's research, or at the way the practice of mathematics is evolving.

Indeed, the Mathematical Sciences Education Board has urged that school mathematics should be grounded in deep mathematical ideas such as change, dimension, and uncertainty, whose power can convey the "vertical continuity" of mathematics from childhood experiences into adult life (Steen 1990). Experience with mathematics that is deeply rooted in fundamental ideas helps people understand and function in the world in which they live.

Children and adults inhabit the same world—of change and chance, of shapes and space. Mathematics helps explain the regularity of seasons, the unpredictability of storms, the form of crystals, and the language of the genetic code. The same powerful ideas that have made mathematics the language of science have also made it the chief tool of manufacturing, quality control, planning, marketing, and other necessities of modern business. By emphasizing the vertical continuity of mathematical ideas, schools can open to children the superhighways of thought that lead from simple experiences to sophisticated applications.

Although dictionaries define mathematics as the science of space and number, mathematics is in fact more aptly characterized as the science of patterns. Some patterns are numerical, others visual; some are symbolic, others organizational; some are static, others dynamic; some are predictable, others chaotic. However they appear, patterns account for the immense power of mathematics as the language of our age.

As the science of patterns, mathematics provides a common framework in which one can analyze many of the conundrums and mysteries that arise in daily life. Measurement pervades mathematics, as do the tools required to make use of measurements—algorithms and computation. In the era of high-powered computers, visualization is also crucial: computer graphics make data come alive and enlist the eye in the mind's search for pattern. Exploration, estimation, classification, and optimization are tools that will be employed (or at least must be understood) by virtually everyone—for personal financial planning, for business decisions, for interpreting public-policy debates.

Recent advances in mathematical research provide yet another potential source of strength for school mathematics (Cipra 1993, 1994). Despite a public perception of mathematics as an ancient discipline in which everything is known, new mathematics—much of it well beyond the ken of any but experts—is continually being created. In recent years there have been major advances in such areas as nonlinear systems (e.g., fractals and chaos), statistics (e.g., evaluation of clinical trials, inference from limited data), and geometry (e.g., knot theory, computer graphics). Yesterday's news involved "trapdoor" codes that made possible public encryption of data with absolute security from unauthorized use. Today's news is created by computational biology, from identifying genes that cause inherited diseases to validating DNA fingerprinting. Most new uses of mathematics hardly make the news but are nonetheless significant and pervasive: algebraic formulas in computer spreadsheets; probabilistic reasoning in lotteries; geometric thinking in architecture and manufacturing; dynamical systems in stock and bond trading.

Mathematics enables people to represent relationships and thus make plans that can be trusted; to classify behavior and thus separate the predictable from the random; to model processes and thus anticipate the consequences of actions. As in earlier centuries when mathematics emerged as the language of science, a different mathematics has now become the language of the technical workforce.

NEWLY USEFUL MATHEMATICS

The science of patterns—as distinct from traditional school mathematics—involves far more than arithmetic, algebra, and geometry. It involves data and measurement, computation and logic, chance and regularities, analogies and similarities, estimates and verification. These attributes of today's mathematics are embedded in the NCTM Standards, just as they are reflected in the mathematics that one finds in life and work. The richness and interconnectedness of mathematics come alive for students primarily through authentic situations in which mathematics arises in natural contexts.

The response of schools to new developments in mathematics should not simply be to add new topics to an old curriculum. A more effective response is to extract from new mathematical results what they reveal about the driving forces for change in the practice of mathematics, as today's frontiers quickly become tomorrow's applications. Surely one of these forces is the impact of computers; another is the renewed emphasis on concrete, useful mathematics; still a third is the increasing role of measurement and data. Demands of quality control reinforce the importance of probability and statistics; the ubiquitous role of robotics, computer graphics, and "virtual reality" simulations provide increased demand for three-dimensional geometry; and the insatiable demand for dealing with data in industries ranging from banking to sports reinforces the importance of algorithms, data analysis, and related mathematics. The primary lesson that schools should put into practice is to focus the curriculum on the forces for change in mathematics, not on the results of these changes. Only in this way can we hope to anticipate changes in the way mathematics will be used by ordinary citizens. Instead of asking, "What's new?" we should ask, "What's newly useful?" Whereas forty years ago only engineers needed to know about digital electronics, now every technician must also. Today's efforts to flatten management structures are changing many blue-collar jobs into "white smock" jobs in which workers are expected to make decisions based on data continuously gathered from the work environment. Typically, these tasks require sophisticated uses of elementary mathematics, not clever uses of higher mathematics. So while the mathematics involved may be classical, the applications will be quite new. The mathematics is concrete, the applications authentic, and the pedagogy active—all ingredients for good instruction. Newly useful mathematics is the right mathematics to add to the school curriculum.

THE CHALLENGE OF STANDARDS

An emphasis on mathematics at work can provide a fresh perspective on the ironic perception held by some academics that implementation of *standards* will result in a watered-down curriculum. Those who express this fear argue that in the name of "mathematics for all," highly talented students are being denied the opportunity to hone advanced skills of reasoning and symbol manipulation. They worry that if the entire curriculum is focused too narrowly on the mathematics deemed useful for all students, then topics whose payoff is years away (e.g., mathematical induction, binomial theorem) will be neglected, and it will be the better students who suffer.

Two responses to this concern can be found in a solid workplace-oriented curriculum. First, work skills require a sophistication and precision that can push even the best students to attain mathematical results well beyond those most students achieve in today's classrooms. As they secure a broad foundation of examples and concrete mathematics, students will build their own connections between mathematics and the world in which they live and work. This grounding in specifics will lead naturally to subsequent generalizations and abstractions. Workplace mathematics mimics good pedagogy by moving from the specific to the general, from the concrete to the abstract.

Second, courses that put mathematics in context improve student motivation, and thereby improve the learning of all mathematical topics. Students will know from their own experiences answers to the question "Where am I going to use this?" Changing from a traditional symbol-intensive curriculum to one embedded in authentic work-related situations will enable faculty to identify and encourage students who reveal nontraditional aptitudes for mathematics. A strong work-oriented curriculum can help schools and teachers break the mold of traditional expectations, thus enabling more students to achieve higher levels of excellence in mathematics.

Fear of a weakened curriculum is not the only concern that must be addressed by advocates of workplace mathematics. Most mathematicians harbor a deep conviction that aesthetics rather than utility should be the fundamental motivation for the study of mathematics. Mathematics exhibits a "remarkable beauty, power, and coherence" that Thurston (1990) compares to "a flight of fancy, but one in which the fanciful turns out to be real." For Thurston, as for many mathematicians, the goal of mathematics education is "to share the delight and the intellectual experience of mathematics—to fly where before we walked." Yet Thurston goes on to argue, as have many before him, that "the aesthetic goals and the utilitarian goals turn out, in the end, to be quite close. Our aesthetic instincts draw us to mathematics of a certain depth and connectivity ... likely to be manifested in other parts of mathematics, science, and the world."

For most students, the beauty of mathematics will be recognized, if at all, by experiencing the power of mathematics at work. If by emphasizing useful mathematics we can make all mathematics more appealing, then more students will learn more mathematics. That pragmatic argument, rather than any claim of philosophical priority, is the fundamental premise of this paper.

> Many mechanical components rely on linkages of rods that form a mechanism that can move in certain ways, partly constrained and partly flexible. Such mechanisms are often used to transfer power to drive machine parts. The bars that drive the wheels on a steam locomotive are a common example, as are the arms that control dentists' drills. In flat linkages, the bars are joined by bolts or slip joints that constrain the motion to two dimensions. In other linkages, some joints rotate around an axis, thus permitting motion in three dimensions.
>
> A flat linkage of four bars in the form of a quadrilateral will flex in two dimensions, but once the lengths of the bars are fixed, only certain motions are possible. One use of such a linkage is to transform circular to oscillating motion. Suppose A is the center of a motor that turns rod AB of a quadrilateral linkage $ABCD$ in a circular motion. Vertex D is also fixed, so as B moves in a circle around point A, C moves along a path determined by the lengths of the four rods. In one particular application, the fixed base AD is five feet and the rotating arm AB is one foot long. What motions are possible for point C for various lengths of rods BC and CD?

Suppose a fifth rod is added, making a flexible pentagon with vertices *ABCDE*. *A* and *E* are located at the centers of motors that turn rods *AB* and *ED* in circular motions. If *AE* is five feet and *AB* and *ED* are one foot each, what happens to point *C* as *AB* and *ED* turn at different rates?

Workplace examples such as linkage problems are quite common. They involve simple mechanisms that turn, slide, and rotate to make things go. Anyone who has tried to clear a paper jam from the inside of a Xerox machine has seen multiple examples of mechanisms at work. Anyone who tries to fix such a machine needs to understand how geometry serves the cause of making things move. Problems of this type exercise geometric thinking in powerful ways. They cry out for physical models, which are easy to build. Many variations are possible, and all have good physical applications. The mathematics involves applications of coordinate geometry as well as trigonometry and simple algebra. Much of the analysis can be carried out by either geometry, analytic geometry, algebra, or spreadsheets.

CONCRETE MATHEMATICS

No doubt about it, mathematics in the schools still suffers from public rejection of the "new math." The image of abstract set theory and spelling drills on the word *commutative* turned much of the public into skeptics of anything "new" in mathematics. Mathematicians, of course, had something quite different in mind—the unifying power of common structure built on clear definitions and careful logic. This Hilbert-Bourbaki vision of mathematics still forms the fundamental karma of mathematics education in the minds of teachers and professors and is passed on in many subtle ways to students.

But mathematics in practice is more unpredictable and far less pretty than is the landscape of logic and deduction. It is rich in data, interspersed with conjecture, dependent on technology, and tied to useful applications. Geometry is used not so much to prove results as for modeling and measurement, primarily in three dimensions. Algebra is used not so much to solve equations but to represent complex relationships in symbolic form. Numbers are used not just to represent quantities but also to calculate tolerances and limit errors. Concrete mathematics, what one might term "postmodern mathematics," emphasizes numbers in context—numbers used with appropriate units of measurement, supported by computer graphics for visualization, and embedded in authentic models.

Whereas "new math" brought forward the power and beauty of general, abstract mathematics, concrete mathematics seeks value in what Karen Uhlenbeck (1994) has termed "the messy, concrete, and specific point of view of possibility and example." Mathematics at work is concrete mathematics. It is spreadsheets and perspective drawings, error analysis and combinatorics.

Open-ended thinking required to diagnose problems or to make decisions relies primarily on the "newly useful" areas of combinatorics, statistics, and geometry. In contrast, algebra and calculus—the dominant features of today's curriculum—are used in the workplace more as tools for calculation than as tools for reasoning.

Concrete mathematics, built on advanced applications of elementary mathematics rather than on elementary applications of advanced mathematics, can yield several important benefits:

- Increased persistence of students in subsequent mathematics courses
- Stronger basis for abstract mathematics in later courses
- Greater appreciation for the utility of mathematics
- Enhanced understanding of the discipline of mathematics

This kind of modeling-based mathematics offers an ideal setting for effective group-oriented pedagogy that is ripe for "what if" analysis and that provides numerous opportunities for students to formulate and analyze their own hypotheses. Moreover, concrete mathematics serves as an effective "bridge to abstract mathematics" (Graham et al. 1989).

Unfortunately, concrete mathematics can easily be misinterpreted as merely the old "general math" in modern disguise. Courses in practical mathematics (e.g., "consumer math," "general math," "shop math") have always been held in low esteem. Too often such courses emphasize easy "cookbook" solutions, attract poorly prepared and unmotivated students, are taught by teachers with little interest in the discipline of mathematics, and provide narrow skills of little benefit beyond classroom exercises. Such courses are rightly disparaged as unproductive dead ends. Instead of expanding students' horizons, they limit choices.

Concrete mathematics is not cookbook mathematics. Concrete mathematics is specific but not narrow, focused but not prescribed. It is found embedded in rich, authentic examples that stimulate students to think mathematically. Like personal anecdotes in politics and specific characters in literature, concrete cases are what one remembers. Concrete, practical mathematics works. The challenge for school mathematics is to make what works respectable.

MATHEMATICS, UNDERSTANDING, AND PROOF

Mathematics as a discipline has a strong internal and intellectual drive, unmotivated by external application. The pendulum between the pure and the applied, between the aesthetic and the utilitarian, between the respectable and the questionable, has swung back and forth for a very long time. Renewed attention to concrete mathematics could be just another swing of the pedagogical pendulum.

The pendulum analogy is a natural consequence of the polarity implied by the dichotomous language of pure or applied, general or specific, abstract or concrete. These dichotomies themselves arise out of an epistemology of mathematics as an objective feature of human knowledge. But mathematics is more than just a body of knowledge, pure or applied. It is a socially constructed system of human understanding that exists as much in the minds of people as in the written record of civilization. Thurston identifies "personal understanding" as the purpose of mathematical knowledge (Thurston 1994); he finds it not just in books and journals, but "embedded in the minds and in the social fabric of the community." Senta Raizen, writing about reforming education for work (Raizen 1989), describes "community memory" as a crucial feature of technicians who work daily on common problems. Both Raizen and Thurston comment, from different fields and different perspectives, on the widespread use of concrete stories, anecdotes, and heuristics as a vehicle for constructing community consensus.

Not everyone would agree fully with Thurston's argument that mathematics is fundamentally about personal understanding. Most people, especially mathematicians and mathematics educators, have come to believe that mathematics is, in its essence, a deductive, axiomatic system based on definition, theorem, and proof. Although philosophers might quibble about whether mathematics in practice is really as formal (or formalizable) as its reputation suggests, most mathematicians and educators agree that what defines mathematics as a discipline is that proof is its standard of truth. The NCTM *Curriculum Standards* (1989) reaffirms this understanding by recommending that all students learn "to recognize valid reasoning and construct simple valid arguments" and that students planning for college learn, in addition, "to recognize and construct proofs by mathematical induction and indirect argument."

There are many reasons for emphasizing proof in school mathematics. Students need to learn that proof is a distinctive part of mathematics; that proof is more than plausibility or confirmation; that among the levels of convincing argument, mathematical proof alone yields certainty; and that proof makes possible lengthy yet convincing chains of logical argument. Workplace mathematics, with its emphasis on the value of presentations of one's conclusions, on the importance of explaining reasoning to coworkers, and on the need for accuracy of results, is ideal for emphasizing the value of rigorous logical argument.

KNOWING VERSUS DOING

The central intellectual issue in mathematics education is not about pedagogy but about content. We know from research what is required for effective teaching: active engagement with mathematics in context, supported by a sense of community that provides necessary meaning and motivation. We know much less about content. What it is that must be learned by all students? Do the particular topics learned really matter all that much? Does learning in context make a difference?

Some answers are easy: We want all students to learn to reason and calculate, to solve problems, and to communicate mathematically. Other answers are more difficult: Do we want all students to be able to prove results? To be able to solve quadratic equations? The key argument is about product (concepts, facts, theorems) versus process (logical reasoning, problem solving, mathematical thinking), and it goes right to the heart of what one believes to be the nature of mathematics (see, e.g., Schoenfeld [1994]).

Product gives rise to "knowing" mathematics, to its definitions, theorems, and proofs. Process gives rise to "doing" mathematics, to its calculations, algorithms, and procedures. Although traditional approaches to mathematics education tend to give students the product rather than the process of mathematical thought, nearly everyone agrees that the ultimate test of mathematical knowledge is the ability to do mathematics. Yet this conventional wisdom—that doing is more important than just knowing—is in many respects a radical position, since it implies that the process of mathematical thinking is far more important than the particular topics that serve as a vehicle for that reasoning. This revolution in perspective is exactly what the Standards hope to accomplish: to ensure that students can do what formerly they had been expected only to know.

The shift from product to process, from knowing to doing, requires that students engage mathematics as a whole, not just as a collection of separate topics, chapters, and formulas. Work-inspired problems reveal more readily than do artificial textbook problems that effective mathematics must be approached holistically, not as an accumulation of bits and pieces of decontextualized knowledge. Although the development of mathematical expertise traditionally has been approached by decomposition into component skills, it is becoming clear that this "one rule at a time" approach does not work (Raizen 1989; Dreyfus 1993–94). "It is harder, not easier, to understand something broken down into all the precise little rules than to grasp it as a whole" (Thurston 1990).

Thurston's advice for effective approaches to mathematics teaching derives from his experience as a mathematician. But his advice is the same as that provided by cognitive science, educational research, and workplace practice: Present mathematics in a way that is "more like the real situations where students will encounter it in their lives—with no guaranteed answer.... It is better to keep interesting unanswered questions and unexplained examples in the air, whether or not students, teachers, or anybody is yet ready to answer them."

Context-based situations such as those that arise in life and work provide an ideal setting to develop this kind of flexible approach to solving problems. Product and process blend in "deliberate ambiguity" (Tall 1991) when they arise from a single context. Students first use mathematics as a tool, and in the "doing" they can be helped to develop the cognitive connections between product and process required to understand mathematics as a discipline. Thus concrete, contextualized mathematics can be especially effective as the glue that binds together in the minds of students the many facets of mathematics.

> Robotic devices are often used where reasons of safety or convenience make it impractical to rely on human control. To estimate its location, a robot measures the angles to various known landmarks and then uses this information with a built-in computer program to calculate its likely position. Since the information available to the robot is often somewhat uncertain (a common circumstance known informally as "noisy data," often due to imprecision in measurements), the robot's position can be identified only within a small region, not at an exact location.
>
> For example, suppose a robot is moving over a flat field to find land mines left over from a war zone. Three distant trees are used as the known landmarks, forming a triangle whose base is 2520 feet long with base angles of 65°30' and 52°40' with the third tree. The field to be searched is approximately rectangular, situated mostly within the triangle formed by the three trees but with some area extending outside the triangle. The sensors on the robot can determine angles to the known landmarks with an accuracy of ± 2°, but they have no way of measuring distance. How accurately can the robot determine its position when it is near the center of the triangle formed by the trees, when it is near one of the trees, and when it is outside the triangle? In which part of the field will the robot's calculations pinpoint its position with greatest accuracy?
>
> A fourth tree is located outside the original triangle at a position that forms angles of 38°50' and 95°10' with the two base trees. If this information were added to the robot's "known landmark" data file, how much more accurate would that make its estimation of position?
>
> *This is a very complex problem using routine trigonometry, but with realistic (not oversimplified) data. The narrative form of the question requires a narrative type of answer since it asks for general statements, not just specific calculations. The problem requires some amount of interpretation and judgment and gets into common yet difficult areas (due to overdetermined and possibly contradictory data) about which students may not have had any experience. A computer (even just a spreadsheet) could be used to help organize the calculations, as could a system of complex simultaneous equations. (In real situations, this problem has a tougher twist—not just to find the data but to do it so quickly that the robot will instantly know its current position.*

That is a more advanced question about computer algorithms that requires linear algebra.)

STANDARDS AND EXPECTATIONS

The expressed purpose of the NCTM *Curriculum and Evaluation Standards for School Mathematics* (1989) is to define a mathematics curriculum that will prepare students for lifelong learning, produce a mathematically literate workforce, provide equal opportunity for all students, and educate an informed electorate. These objectives are encapsulated in five goals: solving problems, reasoning mathematically, building confidence, communicating mathematically, and valuing mathematics. This view of school mathematics balances affective, behavioral, and cognitive goals; is based on the premise that "knowing" mathematics is "doing" mathematics; recognizes technology as a major factor in the changing practice of mathematics; and promotes active learning that engages students in constructing their own personal mathematical knowledge.

The Standards have become the mantra of reform in mathematics education, even among many who have not studied them in any detail. For some, the Standards represent more a metaphor than a revolution—a raising of the bar rather than a change in the game plan (or of the game itself). For others they represent a retreat, since the goals of the Standards—but not necessarily the curricular content—appear less focused than those of the traditional "rigorous" college-prep program. Still others see the Standards as a threat to local school autonomy, since they could be viewed as the first wave of legislatively imposed control over school curricula. For others, mostly advocates of the standards-based reform movement, the Standards represent the framework for a renaissance in mathematics education.

Figure 1 provides a sampler of the content expectations of the NCTM *Curriculum and Evaluation Standards,* illustrating both the core curriculum for all students and the enhancements for college-intending students. These expectations reveal certain important features:

- The greater emphasis on breadth (e.g., statistics, discrete mathematics, technology) is a considerable improvement over the traditional curriculum, with its relatively narrow focus on topics required to prepare students for calculus.
- The presence in the core curriculum of topics that will be used only by those who go to college (e.g., operating on matrices, analyzing Euclidean transformations) reveals a disposition that the best mathematics for all students is college-preparatory mathematics.
- The lack of emphasis on sophisticated applications of elementary mathematics (e.g., three-dimensional geometry, measurement and transfor-

Figure 1. A Sampler of Standards
(from the NCTM *Curriculum and Evaluation Standards for School Mathematics*)

For All Students

- Make and test conjectures, judge the validity of arguments, and construct simple valid arguments.
- Translate among tabular, symbolic, and graphical representations of functions.
- Apply trigonometry to problem situations involving triangles.
- Recognize equivalent representations of the same concept.
- Operate on expressions and matrices, and solve equations and inequalities.
- Model real-world phenomena with a variety of functions.
- Investigate limiting processes by examining infinite sequences and series and areas under curves.
- Understand sampling and recognize its role in statistical claims.
- Represent and analyze finite graphs using matrices.
- Compare and contrast the structural characteristics of the real number system and its various subsystems.

For College-Intending Students

- Construct proofs for mathematical assertions, including indirect proofs and proofs by mathematical induction.
- Demonstrate technical facility with algebraic transformations, including techniques based on the theory of equations.
- Understand operations on, and the general properties and behavior of, classes of functions.
- Solve trigonometric equations and verify trigonometric identities.
- Develop an understanding of an axiomatic system through investigating and comparing various geometries.
- Understand the conceptual foundations of limit, area under a curve, rate of change, and the slope of a tangent line and their application in other disciplines.
- Analyze graphs of polynomial, rational, radical, and transcendental functions.
- Test hypotheses using appropriate statistics.
- Represent and solve problems using linear programming and difference equations.
- Prove elementary theorems within various mathematical structures such as groups and fields.
- Develop an understanding of the nature and purpose of axiomatic systems.

mations of scale, authentic multistep problems) suggests inattention to the needs of the workplace.

For contrast, figure 2 contains a sample of standards that might be added to reinforce the need for more concrete, useful mathematics in the core curriculum for all students.

• •

Figure 2. A Sampler of Standards for Workforce Preparation

- Use appropriate tools to measure real objects.
- Determine what measurements are required to accomplish a task.
- Draw, label, and interpret scale diagrams.
- Measure, calculate, and convert using different units.
- Express and predict precision with appropriate tolerances.
- Measure and calculate dimensions of composite three-dimensional figures.

- Use appropriate formulas to calculate indirect measurements in three-dimensional figures.
- Use triangle trigonometry in three dimensions with diverse units of measurement.
- Execute sequential unit conversions in order to put problems in common terms.
- Solve problems requiring multiple calculations with various units.
- Check calculations in several ways to ensure no chance of error.

• •

PROVIDING FLEXIBILITY

Despite many obvious differences between the reality in today's classrooms and the vision of the NCTM *Curriculum and Evaluation Standards* (1989), both reflect the common heritage of a "gene pool" of goals, topics, examples, and approaches drawn from academic experiences, primarily those deemed appropriate as preparation for college. Before the Standards, colleges set the de facto targets for school mathematics, and the whole curriculum fell into line: Arithmetic led to algebra, which led to calculus, which led to college-level mathematics. The alphabet soup of college entrance hurdles (ETS, SAT, AP) served well enough as national standards for schools and students.

With the publication of the *Curriculum and Evaluation Standards* came a significant philosophical and practical shift in the balance of leadership. No longer are colleges in the driver's seat. Now schoolteachers and mathematics educators are calling the shots with a widely heralded report, rooted in research and reflecting best practice. But the Standards, like what they hope to replace, still respond to the siren call of traditional calculus-prep mathematics. In the *Standards,* high school students are identified as "college-intending" or not; the core curriculum for all students is to be the same for eleven grades; and that core is the curriculum taken by those identified as college bound.

The motives for this recommendation are laudable and the logic nearly inexorable: To assure that all students will have equal opportunity to benefit from the power of mathematics, all must receive the same education, an education that had previously been reserved for the elite. Yet three out of four students move into the workforce along a path that does not depend on a four-year degree. Although some form of postsecondary education is likely to be required for any career in the future, for most students the mathematics they study will be very different from the traditional college mathematics curriculum and may not involve any calculus at all. Thus we can no longer assume that students' needs for mathematics education are best met by the first half or three-quarters of a calculus-prep program, old or new.

A major objective of the Standards is that all students have equal opportunity to engage significant mathematics. Many parents and educational leaders are concerned that in the rush to eliminate a dual-track educational system that gave the least help to those most in need, some schools have eliminated honors programs and college-prep tracks that many consider absolutely necessary for the most able students. They also fear that the floor defined by the Standards will become a ceiling. Although the *Standards* does argue that current tracking systems perpetuate serious equity problems by denying students access to important mathematics, nothing in the *Standards* requires restrictions on course options for students with diverse needs (Romberg 1994) or with significant differences in individual knowledge (Usiskin 1994).

Standards mean comparable opportunity to learn significant mathematics. They do not mean one level for all nor the same speed for all. They do mean that each student should be challenged, and none bored. There should be appropriate opportunities for enrichment for those who want it, for support for those who need it, and for flexible transitions so that students can change from one emphasis to another as their interests, motivation, and career goals change. A strong curriculum rooted in the concrete mathematics that arises from the world of work can give all students the flexibility to explore school-to-work as well as school-to-college options, without foreclosing one at the expense of the other.

ADDRESSING COMMUNITY CONCERNS

Since publication of the NTCM *Curriculum and Evaluation Standards* (1989), many in the mathematics community have raised concerns either about the Standards themselves (both what is said and what is not said) or about attempts at interpreting and adapting the Standards. Many of these concerns have been sparked by reports of what one might term "excesses of implementation," such as no tracking, no symbols, no proofs, or no calculation. Mathematics at work provides a new perspective from which to view some common community concerns.

Symbols

The use of symbols virtually defines mathematics as a discipline. Symbols are a tool for thinking and for reading and recording technical material. Through widespread use of spreadsheets, symbolic expressions have also become the language of technology at work, both in the office and on the factory floor. Unfortunately, symbols in school mathematics so often appear only in artificial contexts that teachers find it difficult to motivate students; as a result, teachers and students alike often become symbol avoiders. Workplace mathematics can help teachers and curriculum developers distinguish between the essential role of symbols as part of mathematical literacy (e.g., in reading technical manuals) and their esoteric use in manipulating algebraic expressions, which has much more limited utility for most students.

Algebra

Many are worried that in the name of "algebra for all," students are being required to take a traditional algebra course that is neither pedagogically sound nor practically effective (Silver 1994; Lacampagne forthcoming) because it consists essentially of a list of loosely related topics. Mathematics in life and work, however, does not come equipped with course labels or lists of topics but with free-form problems that draw on all parts of mathematics. So any curriculum that reflects workplace-oriented practice would automatically have a strong incentive to revise substantially both the content of algebra and the way it is taught—in particular, to emphasize algebraic thinking rather than algebraic manipulation.

Geometry

Observations of the mathematics used in work and life reveal that geometry is useful to a large fraction of the population and algebra only to a small fraction. Concrete mathematics, being primarily about real objects, requires a considerable emphasis on three-dimensional geometry and triangle trigonometry. In problem-solving situations, too, geometry seems to be the more valuable tool for diagnosis and interpretation: Visualization and sketching are more useful than solving equations when faced with most real-world problems. Workplace mathematics can go a long way to restore a better balance between geometry and algebra in the school curriculum.

Technology

Nothing has changed our lives more than technology, and the pace of change shows little sign of abating. Technology is embedded in work and life—in calculators, in computers, and especially in computer-controlled equipment. To be prepared for the world of the twenty-first century, students must learn to function confidently in a symbiotic relation with technology from their earliest years. The dual workplace demands of technological literacy and accurate calculation can reinforce in the schools a healthy balance in an area that is especially contentious in some public and university circles.

EXPECTED OUTCOMES

A plumber is not a failed engineer. Going to work should not be interpreted as a failure to go to college. Nor are practical courses necessarily lesser versions of academic courses. Both can provide challenge, excitement, and significant education; both can serve students' search for vocation with dignity and respect. In shifting educational priorities from *some* students to *all* students, the *Standards* implicitly accepts the challenge of educating prospective plumbers as well as future engineers.

A program of study tilted in the direction of concrete, workforce-based mathematics will encourage students to learn differently and to learn different things than does the mainstream college-prep program of the NCTM *Curriculum and Evaluation Standards*. There will be enhanced emphasis both on practical topics (e.g., triangle trigonometry, ratio and percentage calculations, measurement and units, three-dimensional geometry) and on the habits of mind that are required for success in today's workplace. These include the goals of the NCTM *Standards* (communication, reasoning, problem solving) and, in addition, the ability—

- to take initiative and accept responsibility for one's work;
- to cooperate with others and work in groups;
- to plan and evaluate one's own work and the work of others;
- to work with persons of different backgrounds and cultures;
- to make informed decisions based on careful investigation;
- to identify resources necessary to solve a problem;
- to continue learning technically sophisticated ideas and skills.

Procedures and Skills

Without doubt, most of the public—including most parents and teachers—believe that the primary goal for school mathematics is a set of procedures and skills (e.g., division, percentages, fractions, equations) that should be part of every adult's personal repertoire of basic abilities. Business executives and college faculty expect somewhat more sophisticated skills. For the former it may be constructing a spreadsheet and converting measurements from mechanical drawings; for the latter it is likely to involve relationships among binomial coefficients or characteristics of the normal distribution. Students who undertake a strong program emphasizing concrete mathematics will have considerable opportunity to hone the skills they are most likely to use and also will learn to employ resources (charts, tables, books, computers, other people) to obtain information necessary to complete a task.

Proficiency in Problem Solving

Although the rhetoric of "problem solving" permeates NCTM documents, there is little agreement about what the term means or what scale can be used to determine progress toward meeting this goal. Some have in mind the canonical "set problems" of mathematics exams—both word problems and calculation exercises. Others talk about ambiguous, open-ended modeling types of problems based on real-life situations, suitable for group work and for projects that may last days or weeks. A workplace-oriented program would provide students with innumerable experiences in problem solving of all kinds, from the very specific (set a lathe to grind a wheel to specified tolerance) to the very general (determine how to improve the efficiency of the obstetrics ward in a hospital).

Preparation for Postsecondary Education

Because the final years of high school mathematics generally are viewed as a prerequisite for careers in science and engineering, college and university faculty have come to expect that the best students—the ones they hope to see in their classes—will enter college having mastered a robust collection of techniques carefully selected to prepare students for calculus or comparable courses. Today, however, students enter postsecondary education with more diverse preparation, motivation, and career goals. Instead of a dichotomy—college-intending or not—today's reality is more like a spectrum, with a strong workplace-oriented school curriculum fitting comfortably in the middle of the range of student options. The reality of postsecondary education is that most adults will move back and forth between work and education for two or three decades after they graduate from high school.

Preparation for Life and Citizenship

Mathematics and statistical thinking permeate life. The daily news is full of analyses of health insurance, arguments about the risks of secondhand smoke, and reports of new medical treatments. Economists inundate homeowners and business leaders with reports of mortgage rates, tax issues, and other financial matters. Journalists and politicians talk constantly about opinion polls. And of course the sports pages are full of data and statistics. The traditional curriculum, with its goal of laying the foundation for calculus, often fails to provide much that students find of immediate utility. Students who study a curriculum that emphasizes concrete, applicable topics will be better prepared to make rational decisions, whether about planning their work schedule or choosing a home mortgage, remodeling a kitchen or analyzing political speeches.

CHALLENGES

The need to connect mathematics in school with mathematics at work is generated by the interests of students and the needs of society. However, since today's workplace is so different from yesterday's and is changing so quickly, no one can predict what specific mathematics students will need for the work they will do during their lives. Schools too are changing, some with the purpose of interpreting and implementing the NCTM Standards, some to focus on "education for employment," "tech-prep," or "school to work" programs, and some to restructure their entire organization. Examples of "best practice" can be found in institutions throughout the United States, but the variety of approaches is as great as the diversity of the nation itself.

One can see in innovative programs many clues to the effectiveness of curricula and pedagogy that anchor mathematics in the world of work. Nevertheless, many issues such as the following remain to be monitored as more "newly useful" mathematics is introduced into the mainstream curriculum:

1. What are the real mathematical skills needed for today's workplace? How are they likely to change in coming years?
2. Are the mathematical skills required for work significantly different from those required for postsecondary education?
3. How will college expectations change in response to changes in the preparation of students or the expectations of employers?
4. Will a deep understanding of concrete mathematics lead naturally to the ability to generalize and abstract?
5. Can a curriculum built on the needs of the workplace successfully address the broad expectations for school mathematics?
6. Will mathematics learned in one area of application transfer readily to another?
7. What mathematical skills are especially suited to supporting students' learning in science and technology?

These issues, and many others, will require careful investigation as schools move forward to provide students with mathematical preparation suitable both for work and further education.

The premises of this paper are that school should prepare students for work, that students' experiences in school should reflect the world of work, and that a strong program that prepares students for work will also equip them for success in college. These are the standards against which a school should be judged and that it must use for its own self-assessment: to provide a coherent program in which work is a natural extension of study and school is a natural foundation for work. Concrete mathematics will thrive in such a program, and so will students.

. . . .

BIBLIOGRAPHY

Cipra, Barry. *What's Happening in the Mathematical Sciences.* Providence, R.I.: American Mathematical Society, 1993, 1994.

Dreyfus, Hubert L. "What Computers Still Can't Do." *Phi Beta Kappa Key Reporter.* (Winter 1993–94).

Graham, Ronald L., Donald E. Knuth, and Oren Patashnik. *Concrete Mathematics.* Reading, Mass.: Addison-Wesley Publishing Co., 1989.

Lacampagne, Carole. *Report of Colloquium on "Algebra for All."* Washington, D.C.: U.S. Department of Education, forthcoming.

National Council of Teachers of Mathematics. *Curriculum and Evaluation Standards for School Mathematics.* Reston, Va.: The Council, 1989.

Raizen. Senta. *Reforming Education for Work: A Cognitive Science Perspective.* Berkeley, Calif.: National Center for Research in Vocational Education, 1989.

Reich, Robert. *The Work of Nations: Preparing Ourselves for 21st Century Capitalism.* New York: Vintage Press, 1992.

Romberg, Tom. "NCTM's Standards: Linkages to the Past, Present, and Future." Paper presented at NCTM Annual Meeting, Indianapolis, Ind., April 1994.

Schoenfeld, Alan. "What Do We Know about Mathematics Curricula?" *Journal of Mathematical Behavior* 13 (1994): 55–80.

Silver, Ed. "Dilemmas of Mathematics Instructional Reform in the Middle Grades: The Case of Algebra." Quasar Occasional Paper. Pittsburgh: Project Quasar, May 1994.

Steen, Lynn Arthur. *On the Shoulders of Giants: New Approaches to Numeracy.* Washington, D.C.: National Academy Press, 1990.

Stewart, Donald. "Report from the President." The College Board Annual Report, 1992–93. New York: The College Board, 1994.

Tall, David, ed. *Advanced Mathematical Thinking.* Dordrecht, Netherlands: Kluwer Academic Publishers, 1991.

Thurston, William P. "Mathematical Education." *Notices of the American Mathematical Society* 37 (7) (September 1990): 844–50.

____ "Proof and Progress in Mathematics." *Bulletin of the American Mathematical Society* 30 (2) (April 1994): 161–77.

Uhlenbeck, Karen. "Response to `Theoretical Mathematics.' "*Bulletin of the American Mathematical Society* 30 (2) (April 1994): 201–2.

Usiskin, Zal. "Individual Differences in the Teaching and Learning of Mathematics." *University of Chicago School Mathematics Project Newsletter* (Winter 1994).

Changing the Mathematics We Teach

Cathy Seeley

WE CANNOT teach all the mathematics we have been trying to teach. State mandates are too long; district curriculum guides contain too many pages; school expectations include too many old and new objectives; teachers try too hard and too unsuccessfully to do all of what they feel they are responsible to teach; and slow-moving reform in testing keeps everyone focused on what used to be important.

Meanwhile, the world is changing so drastically and so rapidly that we can see that students need to know more than ever before. Sophisticated calculators and increasing availability of computers make obsolete much of what used to be important in mathematics. Grant Wiggins (1989) has suggested that, across the curriculum, it is futile to try to teach all the important things students will need to know by the time they leave school. We must shift from sound bites to substance if we are to adequately prepare students for the world they face.

Mathematics educators used to be able to justify spending the majority of mathematics instruction on arithmetic at grades K–8 and on algebraic procedures at high school, because people in the world needed to do these things either as consumers, as workers, or as professionals in fields using advanced mathematics. In the face of dramatic changes in all of these areas, we can no longer justify this kind of expenditure when students can look around them to see other people outside of school using technology to do the procedures they are learning in school with pencil and paper. And because of both workplace needs and moral awareness, we especially cannot afford to identify a large

Cathy Seeley is director of frameworks and professional development for the Texas Statewide Systemic Initiative in Mathematics and Science. She has facilitated change at the local, state, and national levels, working as a mathematics teacher, supervisor, university instructor, and director of mathematics for the Texas Education Agency.

portion of our high school population as incapable of academic mathematics, destined to repeat one more time their previous nine or more unsuccessful years of arithmetic, this time disguised as "consumer applications" or "life skills." It has become clear to teachers and policymakers alike that such courses are often more a life sentence than preparation for life.

The result of this insight, at least by some policy-makers, has been to mandate solutions like:

- adopting new textbooks,
- eliminating tracking,
- creating *new* alternative courses, or
- requiring all students to take algebra and geometry for high school graduation.

When teachers complain that these solutions may not meet the needs of all students, we see yet another round of mandated reactions, such as replacing a one-year course with a two-year course labeled *slow, extended,* or *paced*. We lose sight of the problem by ignoring *what* we are teaching and dealing instead with how it is packaged. By replacing traditional algebra with paced algebra, we may transform a year of boring, abstract, irrelevant algebra into two years of *slow,* boring, abstract, irrelevant algebra. In many schools, we change the appearance of a program by having all students take a course with the same title or by enrolling them in their age-appropriate grade, only to implement a new kind of tracking as we create five levels of algebra "to accommodate student needs" or pull out selected groups of students for remedial instruction. Simplistic solutions like these tend to miss the critical point that we must change the very nature of the mathematics we choose to teach, and we must drastically change our approach to teaching it.

At the same time, although the rhetoric of reform promotes de-emphasizing many skills and procedures, no one has provided real assistance to teachers on how they should make the critical decisions about what to teach and, more important, what to leave out. Many well-intentioned teachers have become frustrated with the time involved in teaching the extended investigations called for today while being pressured to prepare their students for improved test performance on narrow measures that look nothing like those investigations.

THE ROLE OF STANDARDS

The National Council of Teachers of Mathematics (NCTM) has articulated the best collective thinking of the field in the 1989 publication of *Curriculum and Evaluation Standards for School Mathematics*. These Standards foreshadowed the development of standards at every level of the educational system and in other disciplines. NCTM's *Curriculum Standards* has clearly presented the best thinking of the profession in terms of what direction the mathematics curriculum should take. The brilliance and the frustration of the Standards come from the same source. The brilliance is that they provide enough specificity that there is little doubt about what kinds of mathematics are important today, while stopping short of dictating, even by recommendation, exactly what should be done at each grade level. The frustration is that a teacher reading the Standards may agree with their direction and may even buy into the list of what mathematics is to be emphasized and de-emphasized, but may not be able to fit all that is recommended into a real school year at a specific grade level with actual day-to-day pressures and constraints.

What the Standards give us is a vision of mathematics that centers on students developing mathematical thinking and reasoning through rich opportunities to solve problems and understand concepts. The focus of the mathematics curriculum is on developing the processes of problem solving, communicating mathematically, reasoning, and making connections within and outside of mathematics. Content strands in the mathematics curriculum remain important, but content is seen as emerging from in-depth experience with mathematical ideas. Rules, facts, and procedures that are necessary to solve problems become the *result* of students' work, rather than the starting point. Assessment meshes with curriculum and the instructional process, with a shift toward open-ended performance tasks, rather than short-answer or multiple-choice tests, thus supporting changes in mathematics content and instruction, and providing important insights for teachers as they guide student learning.

In using the Standards to guide decisions about the mathematics curriculum, teachers and policymakers know they must agree locally about exactly how programs will shift and how decisions will be made about what to teach. Teachers are learning how to redefine the kinds of tasks students do, making use of tools like hands-on materials, calculators, and computers, so that students develop both thinking skills and content understanding. They know that teaching this kind of mathematics requires more time so that students can reach deeper understandings than ever before, struggling with hard questions and challenging problems. They see the power of students spending more time to develop their own understanding of mathematics, as they learn to think deeply about mathematical ideas and as they apply what they learn to complex situations. Teachers know this cannot be done without making choices and agreeing on priorities. We must all address how students can become more responsible for their learning, how teachers can facilitate student thinking and reasoning without losing coverage of content, and especially how to make difficult decisions about what to teach and what not to teach. Teachers know that they cannot fit everything of importance into their crowded teaching schedules.

IDEAS WHOSE TIME HAS COME AND GONE

Much of the conventional wisdom about mathematics curriculum seems out of place in the world in which we now operate. In particular, developments in technology, cognitive psychology, and mathematical pedagogy pose substantial challenges to two beliefs that have exerted great influence over curriculum development over the years:

- The mathematics curriculum can be defined by a comprehensive sequence of skills and concepts described in a set of specific objectives.
- Mathematical skill mastery is a necessary prerequisite for conceptual understanding.

For decades, these two beliefs have pervaded mathematics curriculum guides and textbooks, and their influence on standardized tests will continue to be felt for years to come. Nonetheless, we now know that these beliefs simply are not true for today's mathematics and today's students.

Obviously it is important for a teacher to be clear about what she or he is going to teach. However, attempts in the past to identify a long list of narrow objectives have led to students developing a view of mathematics as an overwhelming set of discrete skills. This set of necessary skills has expanded as teachers have tried to include growing expectations of administrators and parents and as they have tried to implement a spiral development where skills reappear year after year. Of even greater concern, those skills have been practiced extensively, to the exclusion of more interesting and important conceptual development and meaningful applications.

Our most successful students have come to view mathematics as a guessing game where they pick one of the rules they have memorized to solve a given word problem. Our least successful students have neither memorized the rules (although they certainly have seen them several times) nor have they any idea which rule goes with which word problem.

The vision of mathematics put forth in the NCTM *Standards* looks very different from this picture, with students learning to think, reason, and solve problems. Students grapple with potentially extended problems that require synthesizing and applying a variety of interrelated skills and concepts, using calculators for computation, and using computers or other tools for organizing information and demonstrating or simulating situations. When students construct their own mathematical meaning or procedures, and when they learn a select set of skills based on understanding, they are much more likely to remember those skills and more able to apply them effectively in appropriate situations.

It may once have been necessary to master certain skills before a student could deal with a complex problem in a real context. Otherwise, when it came time to do a calculation, a student without computational facility would have been stuck. That is no longer the case. Humans have access to technological tools specifically designed to eliminate tedious computation, and these tools allow anyone to use her or his thinking and reasoning skills to deal with interesting and complex situations. Further, the idea that students developmentally progress from low-level thinking to higher-level thinking seems to have no basis in reality. Students of various backgrounds seem quite able to grasp sophisticated concepts when given meaningful opportunities. Conceptual understanding and knowing what mathematical ideas apply to a situation have become far more important than mastery of specific skills or facts. Describing that kind of understanding of mathematical ideas leads to a curriculum that looks very different from a list of objectives, perhaps involving a narrative description or sample tasks that represent the kind of thinking to be developed.

Mastery used to mean a high degree of proficiency of a certain skill, with ideal instruction including ongoing practice over time until the student had integrated that skill with some degree of automaticity. Mastery also used to mean a barrier to access for many of our students as we withheld the *good stuff* to be seen only by that select group of students who had mastered certain skills. Some automaticity is still appropriate, but far less than ever before. Those skills that are deemed necessary for automatic recall need no longer be used as a barrier for students who have difficulty with them.

Because of the power of technology as an equalizing tool, we need not withhold the *good stuff* from a large portion of our students simply because they have not mastered the rote computational skills we wish they had. All students now have access to good mathematical tasks that help them learn how to think, regardless of their success or lack of success in previous years. Today, we have redefined mastery to mean the deep understanding that comes from spending extended time investigating and thinking about important mathematical ideas, rather than the detailed accuracy that comes from repeated drill and practice on low-level skills. Today, mastery can open doors, rather than close them.

THE PLACE OF THE WORKPLACE

Preparing students for gainful employment is an important aim of education, both for the sake of students and the benefit of society. Nevertheless, for educators to limit educational goals to the needs of the workplace would be negligent. Other needs must be considered so as not to shortchange our students. In particular, we must prepare students for knowledgeable, literate citizenship, for the possibility of higher education, and for options that offer a rich future.

The definition of literacy as we look to the twenty-first century is drastically different from its previous meaning. Today, gleaning information from a newspaper or television story requires more and more quantitative understanding and reasoning ability than ever before. People need to be able to understand statistical statements regarding median incomes, trends in government spending, or predictions about environmental issues. They need to interpret data presented numerically, graphically, symbolically, or in narrative form. They need to use tools of technology to manipulate and deal with data. Literacy as we approach the twenty-first century includes quantitative thinking at an unprecedented level. Knowledgeable citizenship requires individuals who can reason and make sound decisions based on available information.

More and more students seem to be headed for some kind of training or education after high school. We now know that even many students who drop out of high school return to alternative programs for high-school equivalency certificates, often followed by enrollment in technical schools, two-year colleges, or even four-year institutions. We can no longer identify a student as "non–college bound," therefore not needing academic mathematics. We should consider all students potentially bound for postsecondary education of some kind, and, therefore, all students should have access to a solid preparation in mathematics.

Even workplace needs are changing dramatically. Although some employers continue to ask for arithmetic proficiency as a condition for employment, an increasing number of employers prefer applicants who can think and solve problems even if they use calculators to perform computations. As members of the business community participate in partnership activities with schools, they can become more knowledgeable about what is possible with a shift in school mathematics toward the vision of the *Standards*.

Professional educators can function as true leaders within their communities by not only involving business people and the community, but by educating and informing them about what students can do if given the opportunity. We can offer employers a new picture of a well-prepared high school graduate—a picture of someone who can use mental mathematics, estimate answers to practical problems, use mathematics to model and solve unusual problems, draw logical conclusions, use mathematics in communication, and generally demonstrate mathematical power as described in the NCTM *Standards*. Most employers welcome this kind of student, even if the student may not be proficient in all the traditional arithmetic and algebraic procedures. Most students, with this kind of background, will have many paths available to them as they choose their occupational future.

All of these needs call for a different kind of mathematical preparation than simple arithmetic or even algebraic proficiency. They call for sophisticated thinking and reasoning using all the tools available. Students need to be able to work together, communicate effectively with and without mathematics, reason, think, and solve complex problems. The traditional mathematics curriculum puts such outcomes in a backseat role to arithmetic and algebraic procedures, but this position clearly must be reversed.

TAKING A NEW LOOK AT SOME OLD FAVORITES

Most educators, as well as the broader community, can readily agree that they want students to develop the ability to think, reason, solve problems, communicate with mathematics, and connect mathematics to the world outside of school. At the same time, few responsible citizens would argue that, in order to develop such abilities, students no longer need any of the more typical components of school mathematics like computation or measurement, even though it is becoming increasingly evident that not everything will fit into the mathematics curriculum. Fortunately, no one is suggesting abandonment of mathematics content. On the contrary, content still plays an important role, but the focus and details of that content are indeed changing.

Computation continues to be a mainstay of the curriculum, but in somewhat different ways from its traditional appearance. Mental mathematics and estimation are even more important than ever, and are seen as critical tools in dealing with situations in students' world—both in their current lives as students and in their future lives as adults. The development of computation must be based on solid understanding of the meanings of the operations, and the bulk of practice should come in contextual situations so that students become familiar with what kinds of situations call for multiplication or subtraction, for example.

Much emphasis on fractions and decimals is probably still appropriate, with a focus on how these things relate to each other and what they mean in real situations. Students need to be able to handle estimation and approximation with all kinds of numbers. The use of calculators allows students to tackle problems involving far more sophisticated computations than ever before. Most of all, when computation is constantly related to its applications in other parts of mathematics and to real problems in the world, students develop a sense of mathematics' span of usefulness.

Measurement continues to be an important foundation of mathematics. Students who spend their time actually measuring all kinds of objects and exploring spatial relationships learn to use the skills of measurement in ways that support a variety of applications in mathematics, science, social studies, and everyday life. Through the grades, students can spend their time understanding how to use and relate basic units of length, volume, mass, time, and temperature. They can use computer simulations to explore very large or very small units, or to experiment with how measurement is used in different contexts not traditionally available in school settings.

Within geometry, the emphasis shifts from low-level vocabulary memorization and figure recognition to investigating geometric relationships and developing spatial skills. Using the power of geometric construction software on a computer, students can apply centuries-old construction techniques to discover generalizations that were never before accessible to anyone but mathematicians. The use of geometric proof as the primary vehicle for developing students' reasoning skills can now be expanded to include work throughout the grades on justifying conclusions and generalizations—with operations, numbers, algebra, geometry, or any other part of mathematics.

Algebra is a way of using a specialized form of representation to model a variety of problems in real contexts. Using the table building and graphing capabilities of calculators and computers, students can discover algebraic relationships through numerical explorations. They can learn in an actual problem setting how to use algebraic ideas to model an approach to a problem and find a solution. Using computer spreadsheets with formulas allows students to see how variables and equations are used daily to model actual situations. They can develop a powerful understanding of how quantities relate to each other and how patterns show relationships, beginning with work in the early grades using concrete objects and computer- or calculator-based explorations of numbers and shapes. From this early work with patterns, they build toward the use of the important notion of functions as a way to represent situations.

Many excellent teachers today know they can develop these and other important content proficiencies, while they emphasize the long-term development of in-depth thinking, reasoning, and problem solving. They see the power of investing in deep understandings so that time spent learning important skills later will be far more efficient than when students are simply presented with lists of things to learn, and they know this investment will yield rich returns in the long-term retention of relevant skills and ideas. They know that by emphasizing the use of writing, talking, and reflecting on the mathematics students are learning, and by giving students opportunities to connect what they learn, that students will become mathematically powerful. Excellent teachers also know, however, that if this is going to happen, something has to go from the curriculum.

GETTING CLEAR ABOUT WHAT NOT TO TEACH

What survives the bottom-line test for the most important mathematics to teach? This decision involves not only considering what is important to teach, but perhaps even more important, involves decisions about what not to teach. Every teacher, supervisor, and curriculum developer must continually ask the question, Is what we are about to include a worthwhile investment of students' time? For years, we have made assumptions about students needing certain skills in order to be able to progress from one level to another in mathematics. Most of these assumptions are now arguably moot. For example, we simply do not need to ensure that all third graders master their multiplication facts, that all fourth graders master multiple-digit multiplication, or that all fifth graders master fraction operations.

Some of these algorithms and rules are ready to die a natural death. Others will continue to have a role in the curriculum, but not as rigidly attached to any grade level as previously. For example, students still need to know single-digit multiplication facts to do mental mathematics and estimation. However, there is no reason to strictly require their memorization of these facts at any particular grade level or as a barrier against moving on. Instead, selective skill development and practice can be incorporated as an ongoing minor emphasis, rather than the major emphasis, of the mathematics curriculum. The majority of the curriculum can include extended investigations in which students use skills and concepts to deal with real problems or important ideas.

A few parts of the curriculum can be eliminated by streamlining some of the excess duplication across grade levels (place value), postponing some topics until later grades (fraction operations in the elementary school), and omitting some topics that have been addressed in previous grade levels (whole number computation in the middle school). Other parts of the curriculum can be deleted altogether by a careful consideration of whether they are important for today's needs.

Technology Plays a Major Role

Technology plays a major role in decisions about what to teach and what not to teach, affecting the mathematics curriculum in at least three ways. First, technology increases the importance of some parts of mathematics. This means that certain skills and understandings need to be incorporated into the curriculum or strengthened in the curriculum on the basis of their applicability to technological concepts or procedures. For example, since some calculators do not handle fractions well, knowing the meaning of the fraction bar as division and being proficient at translating between fraction and decimal representations have become especially important in working with problems where fractions are a reasonable model. The whole field of discrete mathematics, for instance, including work on finite procedures, has become important because computers operate in finite steps.

The second and even more important influence of technology on the mathematics curriculum is that some mathematics becomes less important, or totally unnecessary, when technology replaces it. Multiple-digit computational algorithms may fall into this category, since few people outside of school actually perform such algorithms using pencil and paper. An increasing number of educators now suggest that these multiple-digit algorithms need not appear in the curriculum ever again (Leinwand 1994; Burns 1994). These algorithms offer a minimal return on an immense investment, requiring the bulk of instructional time in the traditional curriculum, yet rarely performed without a calculator outside of school.

But the third influence of technology on the curriculum underlies why these tools are critically important in school. It is that some mathematics now becomes possible because technology allows it. Elementary students can explore the mathematics involved in real situations that arise in the course of planning a field trip or dealing with school issues if they are free of computational expectations. Middle school students can work with meaningful problems involving proportionality if they don't have to deal with remembering fraction computation rules and performing the related lengthy calculations. High school students with graphing calculators can use algebra as a practical model for sophisticated situations that involve linear or quadratic relationships.

Students can utilize new strategies like successive approximation or computer simulation to model and solve problems previously inaccessible to them, using mathematics previously above their level. Using calculators and computers, they can explore multiple cases of problems, making generalizations about relationships, properties, and solutions, and checking their hypotheses instantaneously. Opportunities like these emphasize the need to get clear about what *not* to teach, in order to provide time and opportunities to reach for potential we cannot even imagine.

Wrestling with The Hard Decisions

It is absolutely critical for teachers to wrestle with hard decisions about which skills, concepts, or ideas are no longer absolutely necessary in the curriculum at any given grade level. It is equally critical that researchers, employers, and mathematics education experts support teachers by making clear statements about which specific skills are not necessary if students learn to think and solve problems as described in the *Standards*. Teachers must feel comfortable that what they are omitting will not later harm students or reflect badly on teachers in general.

An important exercise that can shed light on the most critical components of the curriculum at a grade level is the "Kids can't" exercise. Working within a grade level and with teachers from other grade levels, teachers ask themselves to complete the following statement often made about why students cannot deal with certain mathematics topics:

"Kids can't do ... (a) if they don't know... (b)"

Example: "Kids can't do *fractions* (a) if they don't know their *multiplication facts* (b)."

Spending a few minutes brainstorming responses, teachers identify as many (a,b) pairs as they can. The *b* responses turn out to be some of our assumed necessary outcomes from previous grades, and therefore prerequisites for the grade level at hand. The bottom-line test for these responses is for teachers to assume opposing positions, arguing for or against each response. Teachers should challenge each other with ideas like these:

- But, what if they had a calculator? Couldn't they do (a) without (b) then?
- What if they worked in small groups—wouldn't it be OK if some students didn't know some of (b) then?
- Is (a) really all that important today anyway?
- Wouldn't that develop naturally as students do ... even if they didn't know it before?

Continuing this *bottom-line* discussion, teachers can argue down to the real essentials. In working with many groups of teachers on this exercise over the last few years, the bottom line remains remarkably consistent across grade levels with only slight modifications. Teachers want students who have number sense, operation sense, reading ability, and experience grappling with important mathematical ideas in occasionally messy problems. The bottom line is that these kinds of deep understandings are rapidly replacing a checklist of skills for teachers who want to help students learn to think mathematically and use mathematics to solve problems in and outside of mathematics. Isn't it obvious, then, that these are the kinds of emphases we should be identifying in the mathematics curriculum?

FOCAL POINTS FOR CURRICULUM REFORM

As we look at redefining the priorities of the mathematics curriculum, the most obvious step for many educators and administrators is curriculum writing. Given that most rewritten mathematics curricula over the years have largely been ignored, we must look beyond such traditional approaches to defining what mathematics we will teach. Several focal points for driving curriculum reform have emerged in recent years:

- Local curriculum writing
- Public input
- State frameworks
- Replacement units
- Teacher autonomy

Curriculum Reform Driven by Local Curriculum Writing

In some school systems, curriculum writing has become more predictable than summer camp—if it's June, it must be time to rewrite the mathematics curriculum. Curriculum writing has become part of the culture of schools, and examples of mathematics curricula resulting from these efforts abound on the shelves of school libraries everywhere. Whether created at the state level, by school systems, or by school faculties, the mathematics curriculum has gone through wave after wave of microscopic evolution, with new objectives, more detailed breakdowns, and slight shifts of the grade level at which a topic is introduced or developed. Looking back over the last twenty years of mathematics curriculum reform, the scope and sequence of topics and objectives have remained remarkably constant.

Now that it has become painfully obvious that substantive change must occur, we search for how to make that change visible in classrooms. In spite of knowing better, we still tend to think of writing curriculum as the way to change what we teach. Unfortunately, we haven't usually changed what we mean by *writing curriculum*. If what we mean is to have teachers and other partners get clear about what their mathematics priorities are and what traditional mathematics they will leave out, and if this clarification is seen as part of a larger, more comprehensive process, then writing curriculum may be a reasonable start to facilitating change in classrooms. If, on the other hand, what we mean is that we will convene a group of teachers to look at lists of fifty to sixty objectives at each grade level or for each course, and have them decide which ones to reword or which three or four they will omit because they are covered at six other grade levels, then we are wasting teachers' most precious commodity—their time.

Curriculum Reform Driven by Public Input

Since school systems and state departments of education are ultimately accountable to the public, it is not surprising that they might choose to let the public carry the major responsibility for identifying what mathematics will be taught. Within recent years, several states have convened blue-ribbon committees, often appointed by the governor, legislature, or state board of education. These committees may play any role from advisory to actually developing lists of desired proficiencies for high school graduates. Often, public hearings are convened, with state or school system officials promising to implement whatever goals are identified in the hearings.

Involving the public so directly in making curriculum decisions can prove risky. At worst, it can dramatically set back reform efforts by locking educators into implementing the public's vision of "the way it used to be." Particularly if the public has not been regularly involved in participating in school programs, they might be unaware of the exciting new directions that are possible, and they might be reluctant to let go of traditional procedures. Further, it is possible that those who choose to or are able to participate in hearings or committees may come from only a small segment of the community, possibly representing special-interest groups or extreme points of view.

However, the public can play an important role in setting educational policy directions. When there are appropriate educational and public awareness programs to inform the community about school programs and directions, members of the public can be strong partners in setting the direction for curriculum. They can provide insight into the needs of the community, while supporting educators in developing programs to meet these needs. When educators listen to these citizens and take the major responsibility for designing curriculum and programs, the best thinking of all groups contributes to a sound program for students.

Curriculum Reform Driven by State Frameworks

Some states minimally regulate local school systems and schools regarding what they teach, serving only to coordinate funding or assist with special programs. Most states, however, require state-level determination of goals, priorities or objectives, often defining a vision of what skills, knowledge, and understanding a graduating high school senior should be able to demonstrate. Unfortunately, many states that impose state-level accountability requirements or adopt instructional materials at the state level have not always coordinated these programs with their stated goals and evolving instructional focus. In some states, the bureaucratic structure itself prevents articulation between efforts to guide curriculum, design accountability measures, fund instructional support, design professional development programs, certify teacher preparation programs, and adopt instructional materials. Fortunately, much of this conflict is beginning to diminish as states adopt a more systemic view of school change. Policymakers increasingly realize the importance of leveraging these various efforts toward common goals.

State-level policymakers have a responsibility both to coordinate these closely related aspects of schooling and to ensure that the integrated system moves in positive directions that reflect the best thinking of the profession, the public, and the state's educators about what will serve students for their future. This means that state policies at the highest level, including accountability systems, certification requirements, and funding decisions, need to reflect those priorities stated in a framework. A state cannot effectively improve student learning in mathematics if its framework presents a beautiful picture of what teaching and learning can look like, but its policies remain anchored in priorities of the past.

A *state framework* is not a well-defined entity. In the past, the word *framework* might have been used in reference to course outlines, grade-level content descriptions, lists of objectives, sample activities, lists of adopted textbooks, state instructional requirements, or some information unique to the state. In today's rapidly changing picture of reform, a comprehensive state framework has come to be viewed as a necessary vehicle for state- and school-level change. A framework now is expected to provide a clear vision for school mathematics (as in other content area frameworks), support in identifying instructional priorities at each grade level or course, descriptions of assessment measures (both for classroom instruction and state-level accountability), guidance for developers on what should be included in instructional materials, assistance on how to incorporate appropriate technology into teaching, and suggestions for professional development. A well-developed state framework should also articulate the vision in such a way that it provides a foundation for informing and involving the public in the educational process.

In 1992, the state of California published its forward thinking *Mathematics Framework for California Public Schools*. This framework has been used as the focal point for mathematics reform in the state, targeting reform in curriculum, instructional materials, assessment, and professional development. The discussion of mathematics and curriculum in the *Framework*, with its focus on unifying themes and big ideas, broke new ground and put forth the next step in the evolving vision of mathematics described in the NCTM *Standards*. Through its work on the *Framework* and since its release, California policymakers and educators have continued to demonstrate how such a document can serve as a catalyst for reform both in the state and outside the state.

Three characteristics increase the likelihood that a framework will be effective:

- The framework should be a moving target.
- The framework should come primarily from knowledgeable practitioners.
- The framework should guide all aspects of the educational process.

The moving target

There is no excuse today for any framework to be viewed as complete. Teachers may rightfully complain that are expected to teach toward a moving target—that as soon as they are comfortable with a new curriculum or can successfully prepare students for an assessment program, the curriculum is revised or the program changes. This is an important consideration, but nevertheless, the target must move. We should be using stated targets to guide our efforts, not to end them when we reach our initial goal.

As the framework development process ends and a framework document is published, the seeds for discontent are already beginning to sprout. Technology, world events, scientific advances, and societal trends cause the world to change so rapidly that the vision that is timely and important today is out of date tomorrow. The necessary speed of change today requires more than we can rightfully demand of teachers and other educators, yet it is indeed necessary. Theodore Sizer (1985) has said, "How do we improve schools? ... Slowly. Very carefully. And all at once."

Today this is more true than ever. It is not enough to *fix* the curriculum and be done with it; rather we must set in motion a process whereby we are constantly fixing it, even if it isn't yet badly broken. We move the target not because we were wrong the first time or because we expect that teachers will fail in implementing the vision. Rather, we move the target because we must. To settle for achievement of yesterday's goals for tomorrow's citizens would be unconscionable. The challenge for architects of a framework is to identify what degree of shift is feasible given the reality of where teachers and schools are. The shift should be as ambitious as possible, with as much state support as necessary to ensure that teachers can reasonably implement what is being called for. The even more important challenge for state-level policymakers is to design accountability measures based on the carefully designed moving target(s) that both support the vision and appropriately measure whether real growth is occurring.

The critical role of knowledgeable practitioners

Input from numerous partners is important in developing a useful and visionary framework. For legitimacy, however, the major players must be practitioners who are not only knowledgeable about current directions within the profession, but also in tune with the realities of schools and classrooms. Classroom teachers and front-line supervisors must be integrally involved in creating a framework. Academic mathematicians and mathematics educators furnish a different perspective that can enhance the theoretical foundations for the mathematical ideas in the framework. Supportive administrators and informed members of the public can offer important considerations from their different perspectives. Everyone involved in the development of a state vision or state framework should have extensive opportunities for professional development on new trends, current research, and innovative ideas. Leadership development of framework architects should be a primary consideration of policymakers. These individuals not only create a direction and a vision, they become the potential leaders for future waves of reform.

The framework as a guide for the system

For a framework to be truly useful, it must serve as the guiding structure for all aspects of the system addressed at the state level. Assessment measures and accountability systems, in particular, must reflect the vision put forth in the framework. Professional development programs coordinated at the state level should address areas of the vision identified by teachers as high priorities. The adoption of instructional materials and technological tools should be guided by the framework. Teacher certification, recertification, and reward programs should also be consistent with the kinds of professional development and instructional shifts called for in the framework. All of these—and perhaps other—components of the educational system should be coordinated or supported by a visionary framework if the system is to make real improvement.

Further, a framework should give assistance to school-based programs and teacher planning for classroom instruction. With the increasing responsibility of teachers for what happens in their classrooms, they need a usable tool. They should be able to learn from a framework, to rely on it for guidance on decisions about curriculum, and to refer to it for information, resources, and even inspiration.

Curriculum Reform Driven by Replacement Units

Recently, the notion of block replacement units is gaining in popularity as a means of both curriculum reform and professional development (NCTM forthcoming). A block replacement unit is an extended unit of study, lasting several days or several weeks, that is intended to replace a block (chunk) of instruction such as a chapter or two in a book or a traditional unit of study. Block replacement units are becoming common in commercial catalogs, as many companies have teamed up with funded projects that involve teachers and other educators in creating such innovative materials.

When used appropriately, block replacement units can serve to define what mathematics is important. If a teacher were to use a year's worth of such units, collectively the units could define the curriculum. Additionally, they provide the advantage of allowing teachers to "try on" the new curriculum, getting a feel for the vision of mathematics presented in the *Standards*. Teachers can gradually make a transition to the different kind of mathematics being promoted, learning the kinds of questions they can ask to facilitate student discourse and trying out new instructional approaches. Further, many teachers desperately want to see and experience the kind of tasks on meaningful mathematical ideas that the *Standards* describes. Using replacement units as a starting point allows teachers to focus their time on planning instruction rather than on developing materials.

Several potential problems should be considered, however, in viewing replacement units as the primary method of curriculum reform. First, the quality of the units themselves determines whether this approach is sound or frivolous. Second, if units are designed as a full-year program, the scope of the overall set of units is critical. A teacher might use several units that comprise a year's study, only to discover that students have missed some important mathematical knowledge. Also, creators of block replacement units may not specify what a unit is to replace or what should be omitted to make room for the replacement unit. While some units are clearly intended to replace specific traditional units (such as fractions), others may involve more than one strand and may not fit neatly into an existing curriculum. How does a teacher know which chapters may be replaced?

If block replacement units are to be used effectively for curriculum reform, they need to be considered as a beginning, not an end. Teachers still need help in clarifying what they will leave out as they implement new kinds of tasks for students. They also need some ownership in the new curriculum, perhaps by selecting which units they will use or working together to develop additional units where they identify a need.

Curriculum Reform Driven by Teacher Autonomy

Defining curriculum teacher by teacher and classroom by classroom puts the responsibility for prioritizing mathematical content in the hands of teachers. On the surface, shifting responsibility for curriculum reform in this way seems like an appropriate step in supporting the professionalism of teachers. In some cases, however, what appears to be giving teachers autonomy in curriculum decisions may simply represent policymakers abdicating their own responsibility for clearly identifying priorities. They may not wish to invest the necessary resources to define the curriculum, or it may be too difficult to try to reach consensus among educators and the public if they hold different views of what is important.

When teachers are truly given autonomy, they are given opportunities to work and plan together at their own grade level and across grade levels—identifying instructional directions, getting clear about what traditional content they will de-emphasize or omit, and selecting or constructing instructional materials. Curriculum decisions are among the many decisions in which teachers should play a leadership role. True autonomy means giving teachers opportunities to reach beyond their classroom and participate in their professional community, not encouraging them to shut the door and do whatever they choose.

THE ROLE OF SCHOOLS, SCHOOL SYSTEMS, AND REGIONAL SUPPORT SYSTEMS

In a major shift away from top-down systems of managing education, the school has become the site of management and the focal point of change. With this shift, teachers are becoming more responsible for decisions about curriculum, materials selection, professional development, and in some cases funding priorities. For many professional teachers, this is both good news and bad news. The good news is that they are personally involved in making decisions they will have to implement. The bad news is that some schools and school systems have used the shift toward site-based management as an excuse to gut central office staffs and supervisory personnel, especially content-area resource people.

As teachers become increasingly responsible for making important decisions in school-based systems, they need a strong support system. Mathematics resource support staff from school systems or regional service centers can serve an important function by providing professional development and access to materials that teachers might not otherwise have time to find themselves. School systems can provide a service by bringing together educators with members of the community and by targeting appropriate resources to teachers who need them. The role of the central office becomes one of support rather than directives. This support role is critical, and school systems—those looking to save money by eliminating content area support positions—place far too heavy a burden on teachers at a time when teachers must play stronger roles than ever before. If teachers are to adequately function as the professionals they are, they need all the help they can get, so that they are able to make the best possible professional decisions.

TEACHERS ARE THE KEY

Changes in the mathematics curriculum are interwoven with shifts in how mathematics is taught, as described in NCTM's *Professional Standards for Teaching Mathematics* (1991). No matter how many waves of reform go by, the essential ingredient for student learning continues to be the nature and quality of teaching that students experience. The emphasis on thinking and reasoning presented in the *Standards* can only be taught using student-centered techniques, involving lots of opportunities for students to use their thinking and reasoning and to talk or write about what they are doing. Mathematics curriculum reform, then, is largely dependent on comprehensive professional development. All teachers need to learn how to select or construct in-depth meaningful tasks, how to shift their efforts toward facilitating rather than presenting, how to use the tools of technology to really allow access to high-level problem solving and conceptual exploration, how to change their assessment strategies to support extended investigations, and how to ask probing questions that push students to clarify and refine their thinking.

The basis for teachers' work must be a deliberate decision about what mathematics we want students to learn. When that is clear, the other pieces can begin to fall into place—accountability measures, support structures, materials selection, and funding priorities. Meanwhile, teachers continue to be the key to successful reform. They need to be equipped with the most knowledge and the best tools possible, and they need the time and opportunity to use their best thinking, so that students can develop the mathematical power that is so essential for their future.

· · · · · · · · · · · · · · · ·

BIBLIOGRAPHY

Burns, Marilyn. "Arithmetic: The Last Holdout." *Phi Delta Kappan* 75 (February 1994): 471–6.

California Department of Education. *Mathematics Framework for California Public Schools: Kindergarten through Grade Twelve.* Sacramento: The Department, 1992.

Leinwand, Steven. "It's Time to Abandon Computational Algorithms." *Education Week* 13 (9 February 1994): 36.

National Council of Teachers of Mathematics. Addenda series for *Professional Standards for Teaching Mathematics.* Reston, Va.: The Council, forthcoming.

——. *Curriculum and Evaluation Standards for School Mathematics.* Reston, Va.: The Council, 1989.

——. *Professional Standards for Teaching Mathematics.* Reston, Va.: The Council, 1991.

Sizer, Theodore. "Common Sense." *Educational Leadership* 42 (March 1985): 21–2.

Wiggins, Grant. "The Futility of Trying to Teach Everything of Importance." *Educational Leadership* 47 (November 1989): 44–51.

Rich Content

Richard P. Mills

"*W*HAT *do you mean, about average?*"
While shadowing a school principal in the waning days of last year, I encountered many classes that resembled hundreds in my own experience—teacher at the front of the room, students listening in rows, mostly passive. But the eighth-grade mathematics class was something else. The uproar in this class bespoke real engagement. The room was still the same—chairs in rows—but the students arrayed in twos, fours, and fives had overridden the structure and were obviously excited about the problems at hand.

Without introduction—everyone was too intent to worry about visitors—I tried my hand at one of the problems and was quickly pulled into the cross-team banter about alternative approaches and plausible solutions.

One student worked alone. "Are you a good math student?" I asked, temporarily out of more sensible gambits. "No," she said, "about average." I pounced on that one. "What do you mean, about average?" She led me out of the class, through a hallway, and into another classroom; ignored the health class under way there; and pointed to a chart on the wall. "Average according to that," she said. She was pointing to the Vermont mathematics assessment scoring rubric. Any mathematics teacher would have detected the NCTM origins of that chart.

That one student signaled the success of the whole school. Somehow, the teachers had enabled her to visualize success. She had internalized a set of standards—both what she should know and be able to do, and how well—and she knew what she had to do differently to perform at a higher level. She could talk thoughtfully and self-confidently about her own understanding and performance. That she would eventually attain a powerful understanding of mathematics seemed assured.

Richard Mills is Commissioner of Education for the state of Vermont. He has worked with a continuously expanding partnership to restructure education for high performance. He serves on the boards of the National Center on Education and the Economy, the New Standards Project, and the National Assessment Governing Board.

Getting thousands of students to make that same connection between content and performance standards and actual student work is a long and messy business. I can't say exactly how it is happening, but I can say that it does *not* happen in the orderly way that we imagine when we outline the elements of systemic education reform.

In Vermont, we began with a few dozen teachers bent on creating a new kind of assessment, infused the NCTM Standards into the scoring system, and worked hard on professional development. Later we began to consider the curriculum framework and content standards, then came back to professional development and kept tinkering. How can all schools find their way to rich content in mathematics? Here are some of the conditions that made it happen in one place.

A JUDGMENT OF QUALITY BASED ON REAL WORK

The story of the Vermont portfolio assessment is told elsewhere, but a few points need to be made here in connection with the search for rich content (Mills 1989; Koretz, Stecher, and Dilbert 1992). The portfolio assessment was developed to yield information straight from the classroom to accomplish two purposes. One was to improve teaching and learning, and the other was to support sensible discussions about performance and accountability. These two purposes often seem in conflict in the view of many observers, but both purposes must be served to make real improvement.

At the heart of the Vermont assessment is the conviction that teachers' judgments about students' work can be trusted. Refining the ability to make these judgments and seeking ever greater agreement among teachers about the meaning of the standards that support those judgments is an extremely time-consuming endeavor. The discussions that ensue among teachers have continual reference to students' work and about what content inspires work of high quality.

Since the work that appears in the portfolio is a sample of the best work done during the year, the teachers are assessing the curriculum that is actually followed, not the one proposed or imagined. Consequently, one gets an immediate view of the quality of the content. As a frequent visitor to classrooms, and one who always reads students' portfolios in the course of those visits, I can say that the content is not always as strong as we want. But there is no hiding the quality.

Given the importance of teachers' judgments and the developmental nature of the portfolio assessment, the technical quality of the assessment is essential. Vermont's annual RAND evaluations ensure repeated public examination of data on reliability. This is not comfortable, but it is unavoidable. As Vermont assessment director Sue Rigney noted, we tend to remember that we worked very hard and we recall the anecdotes that indicate massive change in teaching and learning, but change is not the same as improvement. The annual confrontation with the reliability statistics causes everyone to look again at the work assigned to students, the meaning of the standards, and the way teachers make judgments about the work.

A VIRTUAL FACULTY—TEACHERS AS COLLEAGUES

Rich content in mathematics requires solid teaching, and a particular habit of professional teachers, and indeed, of all professionals. The search for rich content requires that teachers band together. Here's an illustration:

The pilot results from the Vermont portfolio assessment in the initial year looked intriguing, and the press briefing, complete with articulate teachers, could hardly have been better. But the headlines the next morning were devastating: "Math scores don't add up." The pilot had revealed generally low performance in mathematics because the teachers had told the plain truth about what they saw in the portfolios—so the press hammered the schools. The initial teacher reaction was anger and a sense of betrayal. But within weeks a remarkable thing happened. Operating largely on their own, with a recent fourth-grade teacher at the head of it all, teachers created more than a dozen professional development networks. Each one had a teacher-leader. And each fourth- and eighth-grade teacher received a list of free professional development opportunities, a notebook crammed with problem sets, the day and night phone numbers of the local network leader, and a free computer access number courtesy of a local telephone company.

During the next year, teachers drove network use to alarming levels and burned up far more professional development funds than expected. People got on the computer network to share problem sets, ask advice, and trade strategies that worked. In effect, all the fourth- and eighth-grade teachers were behaving like the faculty of a single community. They were becoming a virtual faculty, with shared views on expectations and standards. Through state funding of these networks and careful planning of professional development and scoring sessions, people have worked hard to promote this sense that there is only one faculty and all its members support one another.

I spent an afternoon at one of the network sessions and saw firsthand how difficult it is. We passed around benchmark papers that had been scored by expert teachers and tried to match our scores with those of the experts. We worked in pairs, making judgments, arguing with one another about the values we assigned to papers for each criterion, and often disputing with the absent expert colleagues. This kind of professional disputation is a continuous feature of all the scoring sessions in the summer, and it turns up in hot letters in my mail most weeks. Imprecise and time consuming, sure, but the argument is about standards of content and performance. And not just the standards that appear on the wall charts, but the standards that come alive in teaching practice.

Experienced teachers began to express concern about new colleagues emerging from the colleges. Will they know the standards? They will now. Teacher preparation programs, at the urging of the college presidents and the Professional Standards Board, have determined that all teacher candidates will themselves develop portfolios to document their professional competence and their right to enter the virtual faculty.

The extent of the virtual faculty is far wider than this one state, however. Anyone who has observed the New Standards Project is struck by the power of the personal connections that grow as a result (Mills 1993).

THE ACCESSIBLE REVOLUTION

Revolutions are bewildering to participants as well as onlookers. They start, they sort people out quickly. The revolution in mathematics education appears in a very different light. This has been the most accessible of revolutions. Several years ago I spoke to all the school principals in the state about the handful of booklets and papers that one needed to read to gain entry. The whole set made a modest pile on the lectern, and each paper was brief, as revolutionary pamphlets usually are.

Among the powerful ideas contained in those papers were these: From *Everybody Counts*, we learned that the "half-life of students in the mathematics pipeline is about one year" and that while the nation spends more than ten percent of its education budget on mathematics, we are hemorrhaging talent by losing half the mathematics students each year (National Research Council 1989, p. 7). In *Reshaping School Mathematics*, we were urged to reject the myths—many of us thought they were facts—that mathematics "is a fixed and unchanging body of facts and procedures; and to do mathematics is to calculate answers to set problems using a specific catalog of rehearsed techniques" (National Research Council 1990, p. 4). We read that mathematics is "a science of patterns" and a way to communicate ideas that can only be approximated in another medium.

It seemed urgent at the time that every one of those principals should read the fundamental statements emerging in the late 1980s and 1990s right away, before the pile got too large. For months after that, when I visited schools I would ask the mathematics teachers to tell me about the NCTM Standards and most could not. But in the years since that has all changed. Everyone from the teachers in the mathematics network to Colorado Governor Roy Romer has waved about copies of the *Curriculum and Evaluation Standards* (NCTM

1989), and many teachers apparently consult it frequently. And while time has passed, the revolution has not run away from us.

THE RESPONSIBILITY OF LEADERSHIP

The corollary to the accessible nature of the revolution in school mathematics is that people who do discover what is going on have to do their part. Most principals probably have modest preparation in mathematics. And yet enough of them have dipped into the accessible writings about mathematics to enable them to support teachers through the turmoil of changing assessment, practice, and standards all at the same time. The RAND research reports that in the opinion of teachers, principals gave critical help in the form of released time for training and scoring.

The first of many RAND studies of the Vermont portfolio assessment reported that in half of the schools, principals had extended the assessment to grades beyond the fourth and eighth grades intended by the state (Koretz, Stecher, and Delbert 1992, p. 19). Two years later, RAND reported that the figure was 70 percent (Koretz, Stecher, and Delbert 1994, p. vi). All of that extension beyond the fourth and eighth grades occurred without financial support or even professional development opportunities from the state. The strain on schools must have been enormous.

Of course, not all the leading was done by principals. Many of the leaders came from teachers themselves, teachers who are not just willing to grind out the solutions to tough technical issues but passionately committed to doing it right. The portfolio assessment also requires policy leadership willing to stay with a developmental program. There have been so many moments where key legislators, Governors Madeleine Kunin and Howard Dean, members of the Business Roundtable, and the state board seized opportunities to champion the effort and provide the gift of time to those on the front line.

THE AQUARIUM PROBLEM—AND OTHER PROBLEMS WORTH SOLVING

Leadership, collegial behavior, accessible ideas, clear standards of content and performance—all pave the way to rich content. But we need good math problems. For a while, the lack of good problem sets was commonly lamented. During a visit at one school near the end of the first year of the Vermont assessment, a teacher said that not only had his students blasted through all the fourth-grade problems, they had demolished most of those in the eighth-grade packet as well. Was this bad news? Hardly. Bootlegged problem booklets started to appear like cookbooks at community fund-raisers. Some gifted problem builders emerged, and with them has appeared evidence of exceptional understanding of the nature of mathematics on the part of teachers.

Book day in Ellen Thompson's mixed fourth–fifth-grade class in Colchester, Vermont, is one long problem. She has more than three thousand books crammed into her classroom, and the annual inventory starts with a simple task: count the books. But what is a book? Should we count the occasional periodicals the same way? How should one deal with sets? Can the process be speeded along by estimation? How do we consider multiple copies? Can we divide into teams and share the data? How do we verify the counts? How has the collection grown since last year? What is its volume? Its total weight? Will the floor hold? What is the typical book about? And what does *typical* mean in a mathematical sense? When will we be out of space? A good problem is one of the essentials of the drive to rich mathematical content.

One of the New Standards Project problems stands out as an example. It's called the aquarium problem, and it is one of the best ways to show a lay audience why the emerging mathematics is so exciting. The problem consists of a catalog of fish and a letter to the student requesting a plan for an interesting selection of fish that cost less than $25 and fit within an aquarium of a certain size. The problem elaborates rapidly, however, when the student notes that some fish cannot be grouped with certain other varieties; that others can be purchased only alone—they tend to eat one another—and that still others are to be had in pairs or in small groups. And one cannot purchase more than a certain running length of fish in total. Clearly, there is no one right answer, but there are lots of boring answers as well as answers that exceed the budget, the design limits of the aquarium, or the tolerance of the fish.

There is a growing sophistication among teachers and students about what constitutes a good problem, and there is lively criticism leading to better quality. Eighth-grade mathematics teacher Sandra Neustel wrote an open letter to her colleagues about this at the end of the 1994 scoring session. "Be aware," she wrote, "that not all of tasks in the ... resource book issued by the state several years ago are assessable... [and] while some tasks are scorable, they may not be `rich' tasks. As a state and as teachers, we have developed much better material" (Neustel 1994, p. 1). She goes on to note that some collections of student work are bland because there are too many problems of the same type. Portfolio readers, not to mention children, can stand only so many Venn diagrams.

Some teachers are gaining local distinction as good problem creators, and their work is appearing in New Standards Project materials and elsewhere. No longer is this a task taken on by all teachers as they prepare privately for classes, but it is the work of the most skilled of the profession, and their product is valued and used by the whole profession in their region and often more widely.

None of this should surprise us. The NCTM Standards demand that we become mathematical problem solvers. Our scoring rubrics and benchmark papers lead to careful distinctions about students' ability to define problems and extend them.

DUELING MATH PORTFOLIOS—AND THE IMPORTANCE OF AN AUDIENCE

It doesn't go so well everywhere. There are communities in every state where the demands for traditional arithmetic are insistent and where teachers who are leaders to their peers around the nation are regarded by neighbors as soft on fundamentals. But parents and other members of the community need to see the work of students and comment on its quality.

It's not enough to have a good assessment. We have to invent ways for people to have sensible discussions about student performance. So a superintendent told me, and he was right. As we sketched out a plan for a "school report night" that combined a town meeting with a shareholders report, he agreed to pilot the idea in his town. And when that worked, NYNEX put up a cash award for the first fifty schools to do likewise.

I have attended school report night for the last four years in one town because it is so interesting to watch the explosive growth in skill and the inventive way that this faculty has found to display it to the community. School report night is nothing like the familiar back-to-school night where all the adults squeeze into small chairs for a truncated version of their child's schedule, all the while listening to rehearsed presentations from the principal and teachers. School report night is part town meeting, part trade fair. The center of attention is the work products of students. And the talk is about the standards used to judge that work, about how good it is, and about what we are going to do to make it better next year.

The first Cabot School report night was notable for what I have always remembered as the dueling math portfolios. There were two mathematics classes in a particular grade. It seems that the "upper" class had crafted a problem for the "lower" class to solve as a group. The lower class had solved the problem, modified it to make it far more difficult, and then fired it back to their tormentors. The results of this continuing mathematical dispute were displayed across one whole wall of the gym. There were the charts, tables, computer studies, sketches, estimates, and alternative solutions. And members of the public puzzled over work that they had never expected their children were capable of—work that was in fact beyond the skill of most of the adults. And that became the quality of work that set the standard of performance in Cabot—until next year, that is, when the standard was still more challenging.

A COMMON CORE OF LEARNING

The literature on systemic reform makes it clear that states need to start with standards of content and performance. That's true. But it doesn't happen that way. Schools, states, and nations have to take up the thread that comes to hand. In Vermont, people were ready to think about assessment, so that's where we began. All the other pieces had to fit in as well as they could. But once the assessment was launched and panels of expert teachers had made judgments about what to look for in writing and mathematics, we turned to the much broader topic of what we meant by competence.

A few states, notably Connecticut and Maine, had developed statements of a common core. It seemed reasonable to use that material, assemble two dozen of the best minds, and put together our own version in short order. But others counseled a quite different approach. What evolved was a two-part approach that involved 4000 citizens in writing a short Common Core of Learning and a related effort by the profession to build a curriculum framework directly from that public statement of expectations. In short, we asked the public to define basic expectations and then turned professional matters over to the profession—but did so in a manner utterly consistent with what the public had said.

The Common Core focus forums were gatherings in schools and town halls all around the state. Participants were given three-by-five-inch cards and asked to put in mind a young person who was important to them. What did that person need to know and be able to do to succeed. They thought and wrote, then discussed with one other person, and then pairs turned to each other and continued to discuss. The format was simple in the extreme, although discussion leaders often found it very difficult to halt each round of discussion. Many participants said that they had simply never been asked what they thought students should know.

The yield from each community forum was collected and returned to all participants in typed form. Halfway through the months-long process, the Department of Education published a draft and a request for comments. I had marked up my copy—"Looks good!" "This is jargon," "Do you expect this for all children?" and so on. The drafting committee published it with my scrawled comments to signal the urgent need for reader comment. Within a week, hundreds of the copies we had sent out flowed back covered with handwritten praise, criticism, suggestions, and many questions.

We selected a group of teachers known for their own rigorous standards of practice and asked them to evaluate the criticism and prepare another draft. In the end, we distributed more than 170,000 copies, enough for every household in Vermont.

The Common Core included twenty statements of vital results that fall into four groups—communication, reasoning and problem solving, personal development, and social responsibility. The public further expected these vital results to be worked out in three fields of knowledge—arts and humanities; mathematics, science and technology; and history and social science.

The "reasoning and problem solving" vital results, for example, include the following:

- Asks meaningful questions
- Chooses and uses effective means of solving problems
- Approaches problem solving with an open mind, healthy skepticism, and persistence
- Can apply logical strategies to solve problems
- Can think abstractly and creatively

As the professionals began work on the curriculum framework or content standards, they thought of the vital results and the fields of knowledge as two faces on a cube, and a third represented grade levels. In thinking about the content that we expect students to command, sometimes it is useful to sight along one face of the cube and ask about problem solving in mathematics, the arts. And sometimes we want to know about expectations in science in relation to all the vital results. On still other occasions, we just want to know what's going on in the eighth grade.

As the commissions working on the framework presented their first draft for public comment, they did not describe a series of disconnected frameworks for each of the traditional disciplines but rather an integrated structure of what students should know and be able to do that embraces all the disciplines as well as the real-world skills of communications, problem solving, and the rest.

In another year, a statement of performance standards will emerge from the same kind of process. And everyone knows that the whole enterprise will provoke a new burst of creativity in assessment, professional development, problem design, and practice.

MASTERFUL TEACHING

Of all the factors needed to bring rich content to the schools, none is more important than this last one—the visible presence of masterful teaching. There is so much sentimental hokum written and said about teaching that we often need to visit places to see the real McCoy.

The Pew Forum provided such an opportunity when it gathered a panel of scholars in the summer of 1994 to consider how national standards in all the subject areas might be adopted by the soon-to-be-created National Educational Standards and Improvement Council. But they wanted to start with a simulated mathematics class based on the notion that all children could perform at the level indicated by the national standards.

Jill Rosenblum, a former teacher of the fourth grade and now my colleague at the Vermont Department of Education, and Carol Hasson, a fourth-grade teacher in Stockton, Vermont, were asked to teach this "class." It was unlike anything we had seen. The teachers divided us into teams and gave us the task of describing the "typical member" of the whole class. We had an hour to pose our questions, gather whatever data we thought best, build our charts and other representations, and then offer a three-sentence report.

Some groups did surveys, others used estimates, a few relied on stratified random samples. Our teachers coached, prodded, gently closed off dead ends and opened more promising avenues, and kept the teams lurching toward good mathematics. The results were exciting in their originality and in the sweep of the mathematical thinking that emerged.

Afterward, David Cohen, Howard Gardner, and other forum participants put a series of searching questions to the teachers. As they encountered the standards, what did they have to learn and unlearn? How had their practice changed? And for more than an hour, without notes or hesitation, these two teachers opened the new reality of teaching and learning. When they left, our group was silent for a while, and then someone observed how remarkable it is to encounter really masterful teaching. Our task is to make what these teachers do commonly known and commonly available in schools.

SUMMARY

How do we get rich mathematical content into the classroom? There is no formula likely to work everywhere, but the Vermont experience suggests a number of strategies. For example, it is important to assess quality on the basis of real student work. Such an assessment system yields information straight from the classroom to improve teaching and learning and support public discussion about accountability. Rich content also requires that teachers band together across school and district lines to share strategies and seek consensus on standards of content and performance. Not only teachers but also administrators and others must read the essential documents on the new approach to mathematics. And leaders at every level—teachers, commissioners of education, policymakers, and administrators—must in fact lead. They need to sustain teachers through the challenges of changing assessment, practice, and standards—all at the same time.

Rich content depends on a torrent of really engaging mathematics problems. And student problem-solving efforts deserve a public audience. Educators must engage the public in defining a common core of learning, and then the profession ought to honor that statement of public expectations by using it as the basis for content and performance standards. But most important, rich content emerges through the visible example of masterful teaching.

. . . .

BIBLIOGRAPHY

Koretz, Daniel, Brian Stecher, and Edward Dilbert. *The Evolution of a Portfolio Program: The Impact and Quality of the Vermont Program in Its Second Year (1992–93)*. Draft. Washington, D.C.: RAND Institute on Education and Training, National Center for Research on Evaluation, Standards, and Student Testing, July 1994.

_____. *The Vermont Portfolio Assessment Program: Report on Implementation and Impact*. Draft. Washington, D.C.: RAND, National Center for Research on Evaluation, Standards, and Student Testing. July 1992.

Mills, Richard P. *New Standards Project*. Pittsburgh, Pa.: National Center on Education and the Economy and the Learning Research and Development Center, University of Pittsburgh, 1993.

_____. "Portfolios Capture a Rich Array of Student Performance." *The School Administrator* (December 1989).

National Council of Teachers of Mathematics. *Curriculum and Evaluation Standards for School Mathematics*. Reston, Va.: The Council, 1989.

National Research Council, Mathematical Sciences Education Board. *Everybody Counts: A Report to the Nation on the Future of Mathematics Education*. Washington, D.C.: National Academy Press, 1989.

_____ *Reshaping School Mathematics; A Philosophy and Framework for Curriculum*. Washington, D.C.: National Academy Press, 1990.

Neustel, Sandra. "Open Letter on the Mathematics Portfolio Scoring." July 1994.

Vermont Department of Education. *The Vermont Common Core of Learning*. Montpelier, Vt.: The Department, 1994.

_____. *First Draft Content Standards: A First Step Toward Vermont's Common Core Framework for Curriculum and Assessment*. Montpelier, Vt.: The Department, July 1994.

CONTEXT

To understand the process of educational change and the role of professional leadership, the educator must take into account the complex nature of context. There is the profound influence of the broad social and cultural milieu as well as the dynamic character of societal values, conditions, and interactions. There are the myriad institutions and agencies—governmental, quasi-governmental, and private—that play their varied roles officially or unofficially in making, regulating, directing, and administering policy. There are the needs and requirements of business and industry as they relate to jobs and careers. There is the sometimes elusive but very real influence of the media as they both inform and take a part in forming public opinion. Finally, there is the constituency that perhaps matters the most and yet is most difficult to read: the parents and, in public education, the public at large. This constituency is numerically huge and far from monolithic.

Clearly, these are overlapping categories with often conflicting claims and opinions. Their dynamic nature makes the ever shifting picture more akin to a kaleidoscope than a mosaic.

The diversity of the essays that follow reflects this complexity of context, yet the reader will find many common threads and themes that help to explain the optimism with which today's mathematics teachers and educators view the prospects for positive change.

—Shirley A. Hill

Schooling U.S.A.

John I. Goodlad

POLITICALLY driven school reform eras have been predictably nonproductive. There are at least four major reasons for this. First, they are almost necessarily restricted to the systemics and regularities of schooling and schools and so fail to penetrate the student-teacher interaction where learning occurs. Second, because these eras tend to occur at intervals of a couple of decades or more, and since there is then a new group of actors, failures of the past tend to be forgotten and repeated. Third, the short-term character of political initiatives simply does not fit the long-term character of educational improvement. Fourth, the need to drum up public support tends to create inflated expectations that schools simply cannot meet. The schools are then further blamed for what they cannot do; even progress in what they can do is regarded as failure.

The current era of school reform differs in at least two significant respects. First, it has been sustained through the terms of two presidents into the term of a third—long enough to learn some lessons and make adjustments. Second, there are indications that these adjustments include aligning with constituencies that do connect with the student-teacher relationship. A sensible strategy is that of assessing educator-driven initiatives that have promise of involving teachers in classrooms.

In this regard, the National Council of Teachers of Mathematics was a forerunner in developing and publishing standards for the restructuring and improvement of mathematics education in the educational system of the United States. The concept has caught on; today the development and assessment of standards in all the subject fields of the curriculum is on the front burner of discourse, planning, and action at all levels of the political interest in school improvement. The question of whether this, too, will pass into the museum

John Goodlad is director of the Center for Educational Renewal, University of Washington, and president of the Institute for Educational Inquiry. Currently, he is involved in a comprehensive program of research and development directed to the simultaneous renewal of schooling and teacher education.

of school reform once tried is the obvious one to raise. But more important is the question of what is being learned about how to effect and sustain educational improvement. Is the serious work of educators to improve the conditions of teaching and learning to be taken seriously and vigorously advanced in the political arena? Examples of such are rare.

Political support for sound educational ideas can be the best of news and the worst of news, largely because interest tends to be of short duration and heralded by inflated expectations. When attention shifts to something new, politicians call for educators to get involved. A few years ago, circa 1990, for example, political reform rhetoric endorsed what many educational leaders had recommended for years: the school as the center of change and the empowerment of teachers and principals to effect renewal. These educators still hold to the concepts, but political attention has moved on, leaving no record of seriously attempted implementation. The message to educators is clear: Catch favorable political winds in your sails when they blow but don't count on them to power you to your destination.

MISPLACED EXPECTATIONS

The school reform movement that arose in the late 1950s and subsided in the late 1960s was stimulated in large part by national hysteria, fanned by our political leaders, over the sudden appearance of a Soviet satellite circling the earth. The galvanic response was an upsurge in calls for intensifying the science and mathematics programs of schools so that we would attain and retain technological superiority throughout the world. In recommending a comprehensive high school curriculum for all, the most prominent educational reformer of the time, James B. Conant, nonetheless urged a beefed-up serving of mathematics and the natural sciences for the best and the brightest (Conant 1959). Federal funds poured into an array of curriculum development projects in mathematics, physics, biology, and elementary school science (Goodlad 1964).

School reform was at the center of President Lyndon B. Johnson's Great Society agenda. The monumental Elementary and Secondary Education Act of 1965 put unprecedented amounts of federal dollars into educational research and development. The rhetoric of justification, heard over and over at the White House Conference on Education of the same year, charged our schools with not only the achievement of technological superiority but also the end of poverty and violence in our cities and, indeed, the attainment of world peace.

By the time the sobering implications of failure to achieve what we had been told were our objectives in Vietnam were becoming part of our psyche, the schools had become simply part of a lingering malaise. In the late 1970s, there was no talk of the crusades the schools were to lead but a growing interest in whether and how they could do a better job of teaching the young to read, write, and numerate. In the absence of flagellation and exhortation, there were grassroots stirrings for local school improvements. And the decline in test scores appeared to have bottomed out. By the early 1980s, there were some encouraging gains (Stedman and Kaestle 1985).

Suddenly, in 1983, terrible news was thrust upon us once more. Not a Soviet-made Sputnik circling the globe but a U.S.-made horror story spreading from sea to shining sea. If another nation (such as the U.S.S.R.) were to have imposed upon us the school system we have, our nation would have considered this an act of war (National Committee on Excellence in Education 1983). Former military antagonists, Germany and especially Japan, were now our economic antagonists, winning in the global economy because of the shabby performance of our schools. Literally hundreds of commissioned reports picked up the message, citing the several major research-based studies of the time wherever their data helped to support charges of blame but largely ignoring their recommendations regarding the difficult long-term improvements required (Boyer 1983; Goodlad 1984; Sizer 1984).

Although the front pages of these reports usually noted the general neglect of schools and the low status of teachers as part of the problem, teachers were once again singled out among the culprits. Higher pay and improved conditions were something they would have to earn through the improved academic performance of their students. Recommendations were conspicuously short on how these improvements were to be effected beyond the assumed stimulus of salary schedules geared to teachers' demonstrated merit—the systemics of schooling one more time.

Connecting schools to individual financial well-being sells exceedingly well on the political campaign trail. The 1986 report of the National Governors' Association took full advantage of what many people like to hear in the slogan that was to ring in our ears for several years: Better schools mean better jobs. Although the inflated rhetoric of the 1965 White House Conference on Education was equally misleading, at least it proposed to commandeer our schools for the most uplifting ends of the common good.

By the early 1990s, there were clear signs of this nation beginning to feel good about itself again, in spite of a litany of serious problems requiring attention: violence on city streets, drive-by shootings, widespread break-ins, drug abuse, AIDS, political and corporate corruption, and more. But these increasingly were being seen as problems in their own right and not just another byproduct of poor schools. The proportion of American-made cars on the streets was increasing, foreign makers were manufacturing more here, unemployment was decreasing, worker productivity was improving, and polls showed confidence in the future was on the rise. Instead of meekly accepting the charges of some political and business leaders that a major cause of the nation's difficulties was the result of individual sloth and citizens' unwillingness to shoulder their responsibilities, more and more people began to question the interest of politicians in their constituents and that of corporate executives in both their workers and the consuming public. The issue of merit rewards for meritorious performance was turned back toward the sources from which it had emanated for teachers some years earlier.

Meanwhile, several research reports came up with the stunning conclusion that those ubiquitous test scores did not decline over the long haul after all (Berliner 1993; Bracey 1993; Carson, Huelskamp, and Woodall 1991). And there had been no dip so pronounced as to warrant the repeated charges of failed schools leading us down the road to disaster.

Polls in 1994 showed a reversal in viewpoint from that of five years earlier: The U.S. economy was seen to be at least on balance with that of Japan and destined to surge ahead in the future. And so the year 1994–95 was declared to be "The Year of Schooling U.S.A.," and President Clinton appointed a national commission to plan a series of celebrations, culminating in a White House Conference on Education, the first since 1965.

Of course, at the time of this writing, the preceding sentence is a figment of my imagination. To my knowledge, the media have been silent on the subject of the school's role in this turnaround—and rightly so. To my knowledge, no governor has gained political mileage out of claims for economic and other impact from the school reform initiated in her state—and rightly so. But an educator is justified in feeling somewhat cheated over the absence of credit for schools as the cause of improved times when they are so regularly blamed for causing the bad. Nonetheless, we should strive to resist such self-indulgence. Larry Cuban (1994) hits the shameful subterfuge straight on: We have been subjected to a scam, and a dangerous one at that.

There are two very grave dangers in exhorting the schools to be instruments for what lies beyond their capabilities. Most of the problems that concern us immediately—such as creating new and better jobs, adjusting the value of the dollar against the yen, reducing homicides, protecting workers' rights, and reducing homelessness—offer few or no clues to teaching for those who work in schools. Expecting schools to solve them takes attention from what schools can and should do and brings more blame to the schools when these problems remain unresolved. It is a no-win situation for schools. They are blamed for failing to do what they cannot do and go unrecognized for what they do and unsupported in what they should do.

The situation described is negative for society as well. With schools the wrongly chosen vehicle for economic and social reform, attention is diverted from the policymaking and engineering that most of society's critical problems require. Selecting schools as the all-consuming target, as periodically is the case, tends to misuse schools and misdirect resources. Hence my opening statement: School reform movements tend to be nonproductive. Educators must be cautious, however, in directing admittedly much-deserved blame solely to politicians seeking popular missions and business leaders seeking excuses for their own omissions and commissions. Those educators who went along meekly, uncritically, or willingly early in the 1990s with such blatantly political bandwagons as America 2000 far outnumber those who were publicly critical. Many who were skeptical kept reservations to themselves in reaching out for the funds promised that far exceeded funds provided. Some who chose to go with the flow were less than subtle in admonishing outspoken critics to keep silent, suggesting that they were politically naive.

Educators as a group must share the blame for not vigorously enough espousing an educational mission for our schools and agendas for its advancement. The fact that so many of them, along with large segments of the general public, were either gullible or not sufficiently concerned to speak out is a much more serious indictment of the education our schools provide than is the outnumbering of Chrysler, Ford, and General Motors automobiles by Audis, Hondas, Toyotas, and Volvos in parking lots. Inability to detect when we are being taken in by our leaders—and failure to be outraged and outspoken when we know that we are—denote serious shortcomings in the conduct of our democracy and, therefore, our schools.

There are those in leadership roles who, seriously interested in buttressing the schools in every way possible, despaired of gaining more support for an increasingly beleaguered school system without connecting it to personal and family income. This is a sad commentary on faith in our people and, to the degree this assumption fits reality, on the state of the American democracy.

APPROPRIATE EXPECTATIONS

It was Alexander Pope who in 1725 wrote, in irony, "Blessed is the man who expects nothing, for he shall never be disappointed." The way in which we equate education and schooling and then make schooling instrumental to the fulfillment of our private dreams and wants is more an issue of misplaced than of inflated expectations. The love-hate relationship we have with our schools may be the result of believing in them so much that we are let down when they do not do what we want.

What schools can do (not easily) and should do aligns with the highest expectations we have for civilization: the provision of education that ensures the full enjoyment of liberty in an environment of civility and *civitas*, with *civitas* writ large to include the whole of humankind. (*Civitas* denotes a framework of civil law. The school environment provides a unique opportunity for the young to understand that the exercise of individual freedom is impossible without a common framework of group requirements.)

The challenge is daunting. For most of us, gratification must not be too long delayed or we tend to lose interest or even the will. Perhaps this is why we look to schools to satisfy the more immediate and measurable. Then, on becoming more thoughtful and realistically long term in our expectations, we quickly engage in reductionism to make the goals of schooling "operational" and "behavioral." Soon, we are into surrogates that have little or no connection to the wisdom, decency, honesty, and caring addressed in virtually all the aims of education we like to cite. Not surprisingly, success in school as measured by marks and tests correlates with little except continued success in school.

The educational historian Lawrence Cremin wrote that it is folly to consider educational improvement without including the "cacophony of teaching" that surrounds and extends far beyond our schools (Cremin 1990). Given this context, getting our heads very clear for the part schools are to play in this cacophony is imperative. With only 1 000 to 1 200 hours a year available to them, schools are a lesser player but they need not be the least effective. Without a clear mission, the primary function of schools—a major one even under the best of circumstances—becomes the provision of day care for the young.

From their seventeenth-century beginnings, when householders were taxed to provide schools not needed for themselves, our public schools have been called upon to fulfill the unique mission of enculturating the young in the civil laws of the community and those traits of virtue espoused in common by the world's religions. The abiding theme for nearly three centuries was education for responsibility; liberty depended on it. The theme to parallel and then surpass this mission in the twentieth century is education for self-realization; freedom depends on it. In the current struggle for the soul of our system of public schooling, the latter dominates—not for the principle of inalienable human rights but for *my* rights.

For too many of us, these are not rights that enable us to serve some purpose beyond self-interest. They are seen as my due, not something precious to human existence for which I am caretaker. Education in schools must contribute powerfully to the emancipation from narcissism that protects us from indulging freely in the self-defeating freedoms that endanger our liberty. There is no liberty when freedom goes unchecked. In the words of Vaclav Havel (1994):

> The only real hope of people today is a renewal of our certainty that we are rooted in the Earth and, at the same time, the cosmos. This awareness endows us with the capacity for self-transcendence.... It follows that, in today's multicultural world, the truly reliable path to peaceful coexistence and cre-

ative cooperation must start from what is at the root of all cultures and what lies infinitely deeper in human hearts and minds than political opinions, convictions, antipathies or sympathies. It must be rooted in self-transcendence.

The awareness to which Havel refers does not come easily. It is cultivated through education. Democracy does not come about easily, either. "It is an entrancing and rare contrivance of cultivated imagination" (Barber 1993). Cultivating the imagination is the essence of education. Clearly, education is the condition necessary to both self-transcendence and democracy. No matter what else schools do and how well they do it, they succeed in their mission to the degree that those in their care are learning and exercising the democratic moral arts: of individuality, of community ties, and of the common good (Oakeshott 1991).

THE TWO-PART MISSION OF SCHOOLS

The mission of schools is twofold. No other institution or agency is so charged. Many parents assume the mission; some guard it jealously. But large numbers are too busy or ill equipped to discharge it. The first part is enculturation. In the United States, we assume that this is enculturation into a democracy and, therefore, a society in which "everyone accepts and knows that the others accept the same principles of justice" (Rawls 1971)—and indeed, the same principles of decency, honesty, and civility. Clearly, there must be some education in common, or how else are these principles to be commonly learned? Providing this education is the public purpose of our schools.

The second part is the development of the self—impossible without, and significantly determined by, the culture. "Inasmuch as valuing one's self and one's plans and abilities is a cultural achievement, self-respect presupposes education as initiation into some culture. Valuing as a warranted, critical, cherishing of plans and abilities requires apprehension of a cultural context and mastery of its canons of assessment" (Kerr 1987). Our cultural context is a political and social democracy. The process of self-transcendence referred to earlier is one of expanding personal efficacy and the common good. As with all education, the process is of the self in a cultural context. A significant part of this cultural context is an endangered species—the common school.

In maintaining that the primary *function* of schools is a custodial one—that of providing day care—I arouse considerable puzzlement, anger, or both. Many people counter by immediately telling me about the basic skills or higher-order abilities that schools cultivate. True, but function is what something is used for, and above all else, schools take care of our children when we are otherwise preoccupied. Once upon a time, when we had use for children elsewhere, schooling was for only a few months of the year. Now it is for one hundred eighty days in most states, and we clamor for more. Most of the systemics and regularities of schools address this custodial function: bus transportation, prison-like buildings devoid of surrounding trees, security personnel and systems, egg-crate classrooms, no breakout spaces, hall patrols, didactic methods of instruction, regulatory procedures, and more. Parents tolerate few deviations from these regularities, especially on short notice.

Interest in their children's safety at school ranks high among parents' priorities. Teachers, parents, and students place student misbehavior high among the school's problems. Concern over the curriculum ranks low with parents, and even the quality of teachers and their teaching ranks well below some noneducational matters in their hierarchy of problem areas (Goodlad 1984). The costs of seeking to maintain stability in the nonacademic functions have increased markedly for schools in recent years. Meanwhile, policy-makers and large segments of the public express outrage over the rising costs and lowered returns of teaching our young people to write, read, and numerate. More and more, the success stories reported by philanthropic foundations supporting individual schools are of schools grown efficient in protecting their students from the intrusions of malaise from the immediate community.

Statistics that help explain the increasing dominance of the custodial function over the educational mission of schools are only beginning to impress upon us why schooling today is so profoundly different from schooling yesterday. About 7 percent of households today fit the mode of 1940: working father, mother at home with a couple of children. Nearly half of our young people under the age of eighteen living in a female-headed household (and these number about one in five of the age group) are supported by a family income under $10 000 a year. The violent crime rate among teenagers has nearly doubled in the last five years. The threat of violence dominates the daily lives of large numbers of people, often removing the accessibility of the very facilities created to keep the young off the streets and out of trouble.

The threat to the school's educative mission is clear. And so is the answer. We must address directly, with enlightened social policy and effective social engineering, all of the human service agencies that constitute the ecology of our communities. To continue to blame the school for our malaise and to exhort it to a monumental curative effort is sheer folly—for both our communities as a whole and schools specifically. Most critical is the need for parental support systems for which designs but not financial support are available: parental education, prenatal care, health services, early childhood education, and more. All of these have been rhetorically praised and then filibustered into political and economic closets.

Schools in various parts of the country have sought to maintain their stability through pursuing one or the other of two different courses. Some in immediate danger of being overwhelmed by outside forces have sought help in pushing back the dangers around them—by creating a kind of moat between themselves and the gangs, drugs, and violence of the immediate neighborhood. Others are busily engaged in bringing positive segments of the community into the school and effecting alliances with other human service agencies. Given the relative stability of the school, there is a temptation to make it the center where all services are provided rather than the hub of a community infrastructure. The latter design will work only if the school is converted into a multiagency institution that operates twenty-four hours a day and 365 days a year. The outmoded regularities of schooling simply do not accommodate what is now required to sustain the community infrastructure.

Much of what has been described in preceding paragraphs is not only highly visible but also cuts across ideologies, economic class, ethnicity, and race. It is relatively easy to envision strange bedfellows coming together for the common good. But Benjamin Barber (1993) addresses malaise that endangers the very concept of democracy: "Society undoes each workday what the school tries to do each school day.... We honor ambition, we reward greed, we celebrate materialism, we worship acquisitiveness, we cherish success, and we commercialize the classroom—and then we bark at the young about the gentle art of the spirit." The gentle art of the spirit that fuels democracy is what education in schools presumably is about.

FULFILLING THE EDUCATIVE MISSION

What the foregoing adds up to is that schools cannot fulfill their educational mission unless their custodial function is well attended to. And even with the custodial function attended to, fulfilling the educational mission is a daunting challenge, in large part because the tenets of democracy are poorly attended to in the cultural context. Consequently, the imperative regarding the educational mission, referred to earlier, becomes critical. Educators must be very clear on the educational mission of schools so as not to be distracted by the many instrumentalities thrust upon them, and very able to meet the demands of this mission. Consequently, the education of educators, one of the most neglected of all educational reform domains, becomes critically important.

There is a unique reciprocity here, a kind of compact between the people and those charged with educating the young through the specified years of schooling. This compact must embrace, at a minimum, six conditions: (1) custodial care that ensures each child or youth physical, mental, and emotional safety in the school setting; (2) the status of each student as a worthy person with rights and privileges no less or no greater than those of any other; (3) access to knowledge that is in no way impeded because of attributes irrelevant to learning such as color, religion, ethnicity, or financial capital; (4) the pursuit of an individualism that respects the restraints imposed by the physical environment and the rights of others; (5) respect for those to whom the community has given the risk-laden responsibility of teaching; and (6) teachers representing the best assurance of being fully prepared, committed, and able to fulfill the school's educative mission.

In the interest of economies of space, I shall address the sixth of these (which has the merit of connecting closely with the others). The school's mission requires a molding of the self in a democratic context. A democratic context requires that everyone know and accept the same principles of justice, decency, honesty, and civility. But children are not born with this knowledge and acceptance; they must learn it. A compulsory system of schooling (or its equivalent) exists for this purpose. But the purpose is frustrated in an environment devoid of the conditions necessary to its successful pursuit.

The citizens of the community must fulfill their part of the compact, whether or not they send children to the local schools, by doing everything possible to sustain these conditions. They now sustain an array of public facilities, whether or not they use them, and they are not permitted to use or abuse them as they wish. Nor are they permitted to claim their private use. Public roads are a good example. Our taxes support them; we may not close them off to others; and we must observe certain rules pertaining to their use. Why should our public schools be exempted from this public purpose by those who seek to divert public money to serve private ends?

When the public fulfills its responsibility, we should expect and, indeed, require our schools to fulfill their part of the public purpose by teaching the young the democratic moral arts. This means establishing and maintaining the custodial contingencies that support and reward the exercise of civility and *civitas* even before the underlying principles are understood and accepted. But schools must go far beyond the establishment of these contingencies, or their function never expands beyond the custodial one. They must provide the educational experiences in which the young engage in the lifetime process of self-transcendence, through which they develop the morality of individuality, of mutual accommodation, and of the common good. This may be done in many different settings and ways. But the school's role is unique: It must teach society's "canons of assessment"—the knowing and ways of knowing that humankind has acquired over time for purposes of understanding what it means to be a human being "anchored in the cosmos."

TEACHING

Surely this is the educating for which educators prepare. Surely this is the educating in which educators engage simply for the satisfaction of the doing. I recall the skeptical economist inquiring into my association with teachers in the League of Cooperating Schools (Goodlad 1975), all working overtime to create more compelling learning opportunities for the children in their charge: "Why are they doing it? Will they be paid more?" The only appropriate answer is that of Louis "Satchmo" Armstrong (now playing to the angels) on being asked to explain jazz. "If I have to tell you, you'll never understand."

The beauty of the implied educational engagement is its penultimate character—not the ultimate but close. It is a refining step in that rhythm of learning and artistry (in this case, the art of teaching) that Alfred North Whitehead (1949) referred to as style. In teaching, it is experienced only in association with another or others. For Michael Oakeshott (1991), it was the art of conversation; he left us this lovely paragraph:

> In conversation, "facts" appear only to be resolved once more into the possibilities from which they were made; "certainties" are shown to be combustible, not by being brought in contact with other "certainties" or with doubt, but by being kindled by the presence of ideas of another order; approximations are revealed between notions normally remote from one another. Thoughts of different species take wing and play round one another, responding to each other's movements and provoking one another to fresh exertions.... Conversation is not an enterprise designed to yield an extrinsic profit, a contest where a winner gets a prize, nor is it an activity of exegesis; it is an unrehearsed intellectual adventure.... Properly speaking, it is impossible in the absence of a diversity of voices.

Oakeshott's description is of democracy writ small. Expanded into a metaphor, the human conversation is of democracy writ large.

Teachers must bring to conversations with their students what I shall call "teaching style." Teaching style has two inseparable components that authorize one to assume the risk-laden responsibilities for teaching the young (Bull 1990). Society and teachers have a moral responsibility to ensure that both of these are present in the classroom as the major part of the compact referred to above. The first is the teacher's comfort with the canons of the subject matter of the curriculum, be it biology, history, mathematics, or all of these. By *comfort*, I mean the ability to move readily in and out of it, to connect it to phenomena of a different species as well as to concepts embedded elsewhere in the structure of the discipline. This rarely derives from a few introductory courses in college; it is not guaranteed by degrees and majors. Nonetheless, it is part of what is intended in a general, liberal education.

The second component pertains to one's view of knowledge in general and specific domains of knowledge in particular. History, music, literature, and mathematics have their own distinctive modes. These come together with the principles and concepts of the discipline in pedagogy, the art and science of teaching (Bruner 1960; Shulman 1986). It has been said that teachers must learn all subjects twice—once for themselves and once more for their teaching. Presumably, it is these modes in particular that the conduct of general education commonly does not provide. Consequently, they must become part of the professional education of the teacher. Years ago, in my role as director of the University of Chicago's Center for Teacher Education, I scheduled interviews with colleagues in the arts and sciences to learn when they addressed in their programs the modes of their disciplines. Communicating on this topic proved to be difficult, but the conclusion was that the relevant conversations occurred in doctoral seminars but rarely before. We begin to understand why an undergraduate or even master's-level major in a subject to be learned and taught does not guarantee adequate preparation for teachers.

The pedagogy that combines these components and what is commonly referred to as methods of teaching, whether generic or specific to a given subject, are not the same thing. Much of what I have observed under the label of methods of teaching is more management than pedagogy: how to maintain an orderly group, cue students' attention, maintain an instructional span of control, present a film, and so on. Students preparing to teach want more and more of this (Goodlad 1990); and from the apparent popularity of workshop after workshop (often carrying high enrollment fees) among in-service teachers, it appears that they, too, cannot get enough of this kind of classroom method. But this approach resembles the filling up of a bagful of techniques that might work on some occasion rather than a disciplined grounding in the science and art of teaching that builds over time into teaching style.

PEDAGOGY AND DEMOCRACY

The second part of the school's mission carries one into the canons of assessment that discipline the human conversation and then into the pedagogical requirements of those who take on the role of teacher in classroom conversations. Does this part of the preceding narrative have anything to do with the first part of the school's mission: enculturating the young into our political and social democracy? I believe that the pedagogy of the classroom day after day, in all of the subjects of the curriculum, has more to do with the development of the democratic moral arts than does any didactic teaching of values and virtue. Further, this pedagogy, embedded as it is in the subject matters children and youth are supposed to master, carries with it the political legitimacy of advancing the school's educational responsibility.

Colin Hannaford, director of the Institute of Democracy for Mathematics, Oxford, England, and a teacher of mathematics in England for twenty years, argues the case for his subject. In many ways, mathematics provides the quintessential illustration, in that the conventional view is of truths given and fixed.

There are facts to be memorized, used in computations, and tested. It is common in educational circles to speak of cultivating creativity through the arts and problem solving through the social studies or even history, but of neither as appropriate for mathematics. Mathematics is frozen in time and space; it is to be learned as given, without question. Nonsense. It was not the inability of school children in the United States to compute that aroused the concern of the National Council of Teachers of Mathematics. It was students' revealed inability to move around comfortably in mathematical terrain, to intuit and inquire.

I have permission to quote some highly relevant passages from Hannaford's lecture to the Ecole Supérieure in Paris (Hannaford 1994, pp. 2–3, 12):

> I have noticed that mathematics is taught and learnt according to a definite morality. This morality is itself democratic.... When we teach mathematics to-day, therefore, we instill our pupils, implicitly, with the ideals of democracy.

> When we teach it as their personal adventure of intelligence and discipline, which they have inherited from, and share with, all the people on Earth, then we teach democracy with self-confidence, freedom, respect for others, and generosity.

> But if we teach mathematics as a closed structure: already closed, that is, already finished in a reality beyond the understanding of the majority of people—then we impress our young pupils with a very different paradigm.

> Instead of freedom, we show them limitation. Instead of self-confidence, we teach subordination. Instead of generosity, we insist on obedience.... Mathematics taught like this is totalitarian....

> Mathematics shows people how to agree: how, intelligently and reasonably to explain their viewpoint and position to one another, and to come to an agreement of a solution; or to agree that there is no solution at all.

In this passage, Hannaford argues a case for the pedagogy of mathematics. Elsewhere, he develops an argument, with historical documentation, for the connection between mathematics itself and democracy, a topic beyond the scope of this paper. My argument here, joining at least part of his, is that pedagogy informed by the canons and modes of the domains of knowledge puts one in touch with democracy. School reform movements are predictably nonproductive because they never get to the conditions necessary for such pedagogy to flourish.

For schools to fulfill their democratic public purpose, three sets of conditions must be in place. First, the custodial environment must be a democratic one. Second, teachers must be in command of the necessary democratic pedagogy. Third, a substantial percentage of the citizenry must agree on the public purpose of our schools. The first two, challenging in their own right, are in jeopardy because of the absence of the third. The mission of schooling U.S.A. increasingly is being seen as a matter of private purpose.

Just as physicians must advance the best in medical practice, whatever the social and political context, educators must advance the best in educational practice. Their interest, currently enjoying political support, in embedding standards central to their fields in ongoing classroom practice will come to little so long as a significant portion of the citizenry and its elected representatives see no need for rigorous teacher education programs comparable to those deemed necessary for the major professions.

· · · · · · · · · · · · · · ·

BIBLIOGRAPHY

Barber, Benjamin R. "America Skips School." *Harper's* 286 (November 1993): 44.

Berliner, David C. "Mythology and the American System of Schooling." *Phi Delta Kappan* 74 (April 1993): 632–40.

Boyer, Ernest L. *High School.* New York: Harper & Row, Publishers, 1983.

Bracey, Gerald W. "The Third Bracey Report on the Condition of Public Education." *Phi Delta Kappan* 75 (October 1993): 114–18.

Bruner, Jerome S. *The Process of Education.* New York: Vintage, 1960.

Bull, Barry L. "The Limits of Teacher Professionalism." In *The Moral Dimensions of Teaching,* edited by John I. Goodlad, Kenneth A. Sirotnik, and Roger Soder, pp. 92–98. San Francisco: Jossey-Bass, Publishers, 1990.

Carson, Robert M., Robert M. Huelskamp, and J. D. Woodall. *Perspectives on Education in America.* Third draft. Albuquerque, N.M.: Sandia National Laboratories, May 1991.

Conant, James B. *The American High School Today.* New York: McGraw-Hill, 1959.

Cremin, Lawrence A. *Popular Education and Its Discontents.* New York: Harper & Row, Publishers, 1990.

Cuban, Larry. "The Great School Scam." *Education Week,* 15 June 1994, p. 44.

Goodlad, John I. *The Dynamics of Educational Change.* New York: McGraw-Hill, 1975.

_____. *A Place Called School.* New York: McGraw-Hill, 1984.

_____. *School Curriculum Reform in the United States.* New York: Fund for the Advancement of Education, 1964.

_____. *Teachers for Our Nation's Schools.* San Francisco: Jossey-Bass, Publishers, 1990.

Hannaford, Colin. "Mathematics Is Democracy." Lecture to the Ecole Normale Supérieure, Paris, September 1994.

Havel, Vaclav. Remarks on receiving the Philadelphia Liberty Medal, Philadelphia, 4 July 1994.

Kerr, Donna H. "Authority and Responsibility in Public Schooling." In *The Ecology of School Renewal*, Eighty-sixth Yearbook of the National Society for the Study of Education, edited by John I. Goodlad, pt. 1, p. 23. Chicago: University of Chicago Press, 1987.

National Committee on Excellence in Education. *A Nation at Risk*. Washington, D.C.: Government Printing Office, 1983.

National Governors' Association. *Time for Results*. Washington, D.C.: The Association, 1986.

Oakeshott, Michael. *Rationalism in Politics and Other Essays*. Indianapolis: Liberty Press, 1991.

Pope, Alexander. Letter to William Fortescue, 23 September 1725. In *The Correspondence of Alexander Pope*, edited by E. G. Sherburne, vol. 2, p. 323. Oxford: Clarendon Press, 1956.

Rawls, John. *A Theory of Justice*. Cambridge: Harvard University Press, 1971.

Shulman, Lee. "Those Who Understand: Knowledge Growth in Teaching." *Educational Researcher* 15 (February 1986): 4, 14.

Sizer, Theodore R. *Horace's Compromise*. Boston: Houghton Mifflin Co., 1984.

Stedman, Lawrence, and Carl F. Kaestle. "The Test Score Decline Is Over: Now What?" *Phi Delta Kappan* 67 (November 1985): 204–10.

Whitehead, Alfred North. *The Aims of Education and Other Essays*. New York: Menta Books, 1949.

Standards-Based Reform

Wilmer S. Cody

T HIS essay might well have a subtitle: "On Aligning Policy and Governance to Challenging Subject Matter." Standards-based reform is a new paradigm for thinking about what the central objective of education should be for all students in America and for thinking about how to accomplish that objective. It is an idea growing among a number of policymakers—and those who would influence policy—who are looking at how state and local school systems, the federal government, and national professional organizations can interact to affect what happens in the classroom. Standards-based reform is thus a set of ideas that would improve the accomplishments of schooling·by changing the policies that shape the enterprise.

Standards-based reform is a marriage of ideas from several different intellectual traditions concerning beliefs about (1) the capacity of children to learn and therefore what schools can and should accomplish, (2) how the policy structure of school systems should be framed, and (3) what the prerogatives and responsibilities of policymakers, administrators, and teachers should be in order for schools to accomplish what the new paradigm says schools should and can accomplish. In some forms, it would create a new political structure of elected officials through which citizens express their will.

It is a paradigm in the process of being formed or invented and will most certainly manifest itself in a variety of ways—some more successful than others. If past efforts of policy reform hold true in this case, many efforts labeled *standards-based reform* will be in name only. The capacity of policymakers and practitioners to latch onto a new popular label with little regard to its central premises seems boundless. The continuing support of teachers in the

· · · · · · · · · · · · · · · ·

Wilmer Cody is director of The National Faculty/Southern Region in New Orleans. He was executive director of the National Education Goals Panel and has served as superintendent of Chapel Hill, N.C.; Birmingham, Ala.; Montgomery County, Md.; and the state of Louisiana. He directed the planning of the mathematics content for the 1990 NAEP.

implementation of the math standards by the National Council of Teachers of Mathematics and its support of policies that are aligned with those standards provide a constructive lesson on what is needed to keep policymaking on track.

As presented by its major advocates, as well as this writer, standards-based reform has at least three major principles or objectives:

1. Challenging curriculum content with high performance standards should be taught to, and learned by, all students.
2. The policies and major programs of the "system" of schooling need to be aligned with the teaching and learning of those standards in ways that guide the efforts of all involved to their achievement.
3. The authority and responsibility of policymakers, school administrators, and teachers need to be "restructured" in a way that delegates to teachers the discretion they need to effectively accomplish the achievement of high standards.

WHOSE AGENDA IS IT?

This new paradigm pervades many current major education reform efforts taking place at federal, national, state, and some local levels.

Federal

Since 1990, standards-based systemic reform has captured the support of past and present federal officials. The National Science Foundation (NSF) has been awarding large grants for State Systemic Initiatives in mathematics, science, and technology. Currently twenty-five of the fifty states receive five-year awards of approximately $2 million each year that are matched with equal amounts of state money. In 1993, NSF began a similar program for grants to the twenty-five largest urban school districts, and in 1994, a Rural Systemic Initiative was announced. Other NSF programs that focused in the past on one part of the system such as teacher development have been recast to require grantees to incorporate project activities into some larger systemic plan.

The Goals 2000: Educate America Act presented by the Clinton administration and passed by Congress provides for the formal endorsement of high national and state academic standards and national occupational standards, and it supports state and local district systemic reform that is standards based. The administration bill to reauthorize the Elementary and Secondary Education Act (ESEA) changes that law and the use of its substantial appropriation to conform to the same principles.

The Clinton administration's agenda builds on initiatives started by President Bush and the governors of the states, who in 1989 adopted six National Education Goals. Goal 3, in particular, is premised on the first principle in saying, "By the year 2000, American students will leave grades four, eight, and twelve having demonstrated competency in challenging subject matter including English, mathematics, science, history, and geography." The full text of the goals and other reports make it clear that the governors and the president meant challenging subject matter for *all* students (National Education Goals Panel 1991).

National Professional Organizations

The high-standards movement was national before it was federal. The National Council of Teachers of Mathematics started the movement to define more-challenging content for the education of all students with its publication in 1989 of *Curriculum and Evaluation Standards for School Mathematics*. In 1991, the National Research Council, at the request of the National Science Teachers Association, five other national science organizations, Education Secretary Lamar Alexander, and the cochairs of the National Education Goals Panel, initiated a process to produce standards for science; that effort is still under way. In 1992, coalitions of other organizations began to develop standards in geography, history, English, civics, the arts, and foreign languages through a national consensus process.

From the beginning, these organizations intended to develop national standards. At present, they have financial support from the U.S. Department of Education and several other federal agencies. The final reports of most of the coalitions were to be published in 1994. The federal support for this activity originated in a report of the National Council on Education Standards and Testing created by the National Education Goals Panel. The council was subsequently authorized and funded by Congress. The report endorsed and called for the development of *national* education standards (National Council on Education Standards and Testing 1992).

Policymaker Associations

In recent years, the National Governors' Association and the Education Commission of the States formed committees to study education reform ideas, produced reports with recommendations for state policy that incorporate some of the ideas of the new paradigm, and have provided technical assistance to member states who desire to design an education reform plan for their state (National Governors' Association 1991; Education Commission of the States 1992).

Business Leaders

Numerous business leaders and business organizations have adopted the concepts of standards-based systemic reform within the last few years. A major participant is the Business Roundtable, an association of the CEOs of major U.S. corporations. In 1990, the roundtable adopted and published *The Essential Components of a Successful Education System*. Included in the essential components are ideas of challenging subject matter for all students, reform that aligns state policy in a coherent way, and authority devolved so that school-based staff play a major role in instructional decisions. The publication states, "In September 1989, the Business Roundtable committed to a ten-year effort to work with state policy-makers to restructure state education systems and ensure that all students achieve at high levels" (The Business Roundtable n.d.)

The Business Roundtable has coordinated technical assistance on strategy for standards-based reform to business, education, and civic leaders in a number of states. In more than a few states, it has arranged with consultants to conduct a "gap analysis" or evaluation of the state's school system using the roundtable's nine essential components as criteria.

States

Parts of this new paradigm are being implemented in many states and a comprehensive paradigm is being used in several. California is well into reform that aligns high standards, assessments, teacher education, and student assessment. Standards-based systemic reform that also includes the restructuring of governance is occurring in Kentucky. Vermont and South Carolina have major standards development efforts under way. A state court decision in Alabama on equity and adequacy of schools in the state directs a series of remedies based on the three premises of the new paradigm.

Enthusiasm for this new paradigm is not universal. Many local school board members have been cool to a proposal that would shift some of their authority to the local school level. Their national and state organizations have not particularly been leaders on this issue. Some conservative religious organizations that are active on education policy matters have opposed the state adoption of curriculum standards, advocating, instead, that such decisions should be made at the local district level. Nevertheless, for the first time in many decades, a common view of how policy might make a major difference in teaching and learning is emerging that cuts across party lines and is embraced by many prominent policy, business, and education leaders at national, state, and some local levels.

THREE PARTS OF THE PARADIGM

The three ideas represent major shifts in thinking that has dominated efforts by policymakers to improve education in this century. The ideas are based on a set of new premises about student expectations, education policy, school organization, management, and the governance of public schools. In the pages that follow, each will be described, at least some of its antecedents identified, and then some issues discussed that will need to be addressed if the new paradigm is to be durable and effective.

1. Idea One: Teach challenging content and apply high performance standards to all students.

This objective is based on a major change in the belief about the capacity of children to learn: from a belief that (*a*) accepts individual differences in student ability as substantially immutable by the time kids start school, (*b*) holds learning time as constant and (*c*) varies expectations for each student to a belief that achievement is a function of time and opportunity. Almost all children can learn complex, challenging subject matter; some may take longer than others.

Antecedents

They should. Since the mid-1980s, state and national political and business leaders have raised the same call with growing unanimity. Schools have got to accomplish more with all students. Jobs today and in the future will require more knowledge and skills than those of the past. The economic opportunities for the individual depend more on his or her level of education than ever before. Productivity increases in the future will require a more knowledgeable workforce in the future than was so in the past. The economic strength, and thus the quality of life of all Americans, depends on an educated workforce.

A Nation at Risk, published by the National Commission on Excellence in Education (1983), carried the message that things need to get better for the sake of the economy, a message that has been repeated over and over again for the past ten years. *America's Choice: High Skills or Low Wages*, published by the National Center on Education and the Economy, made the same case, described how some other countries were more successful than the U.S., and then proposed some policy direction for education (National Center on Education and the Economy 1990).

The message has gone beyond the society's leaders; over the last decade, opinion polls have revealed a growing support among the public for challenging "national academic standards" to guide curriculum content and to evaluate schools and students (Elam and Gallup 1991).

They can. It is the conventional wisdom of practitioners, well documented by research, that expectations play an important role in the learning of all students. Recent research further demonstrates that almost all children do engage in higher-order thinking and complex problem solving (Bereiter and Scardamalia 1985).

Evidence has been growing that students aren't learning challenging subject matter because they aren't being taught challenging subject matter. Course-taking patterns reveal that with other factors held constant, students who take challenging academic subjects in high school perform better on the SAT and ACT than students who take less rigorous programs (LaPointe, Askew, and Mead 1992).

There has been valid criticism that the curriculum had been watered down and higher-order-thinking skills de-emphasized. In some manifestations, the "basic skills" reforms framed knowledge and skills into numerous minute "objectives" in the belief that mastery of each part would sum to a whole. It didn't, and what may have raised expectations for some lowered expectations for others.

"Evidence proofs" have come to light in which individual teachers and groups of teachers in schools have demonstrated that academically underprepared and average students as well as advanced children would respond to and could learn challenging subject matter. The well-publicized success of California teacher Jaime Escalante and East Harlem teacher Kay Toliver in teaching students mathematics provide two examples.

The educational achievement of students in some other countries such as Japan is higher than that of students in U.S. schools. Comparative studies have revealed that one reason is that the content of the curriculum in Japan is more challenging (Westbury 1992).

Issues

If a consensus within the profession on curriculum content and high performance standards is desirable, what process will most likely produce that consensus? What needs to be done? How long does it take? When does a consensus exist? The process used by NCTM worked. It involved thousands of mathematics teachers, teacher educators, and mathematicians. It took five years to develop the mathematics standards to the point of adoption and official publication. Equally important, NCTM has continued efforts to build support for its standards and their use by teachers, illustrating that building consensus and making standards a central focus of educational practice in America does not "arrive" with the adoption and publication date.

While the consensus process in mathematics was not free of conflict, such conflict is more pronounced in some of the other disciplines. When the U.S. Department of Education first received a proposal to develop English standards from a joint venture of the National Council of Teachers of English and the International Reading Association, one sticking point was the reluctance of some of the proposal framers to commit to try to reach an agreement on standards, thus foreshadowing the need for a lengthy process.

The history standards project leaders are discovering that it is going to take much longer than they originally planned to produce a consensus on standards. Perhaps that was to be expected. The standards projects that were initiated in 1992 had two-year time schedules for development and adoption, considerably less time than the process followed by NCTM.

How should the public be engaged? Citizens will be heard, especially when the standards developed as guides are incorporated into state frameworks and local curriculum guides and when textbooks reflecting the new standards appear in the schools. What the schools should teach children is an issue on which many citizens have opinions, and those opinions frequently are strongly felt. California and New York learned that some citizens had strong opinions about what American history should be taught. Vermont has carried out a very extensive process of engaging the public through many town meetings in developing its framework. Using other methods, South Carolina and Kentucky have sought widespread public involvement. It would seem politically prudent as well as wise to engage the public long before policymakers are asked to adopt standards.

How will time to learn more-challenging content be found for those students who will need more time? Should more time on core subjects be secured by eliminating the study of other subjects for those students for whom it takes longer to master more challenging subject matter? At the local school and perhaps at the district and state policy level, the issue will arise of whether or not to prioritize curriculum subjects for individual students. An elementary school curriculum may contain as many as ten or twelve different "subjects," with less than half the day devoted to core academic subjects. Teaching reading through the arts, physical education, health, and so on, is a popular idea, but there is little evidence that it is executed effectively.

Are there practical limits to how high performance standards can be raised? Policy coherence that focuses current resources more productively on challenging content will undoubtedly make "raising the bar" feasible. Major budget increases, however, will likely be slow to come. One-on-one tutoring three hours a day will likely result in very high achievement, but it is doubtful that the public will support such an investment for many students. The press for a longer school year has not been well received by state legislators when they made budget decisions. One feature of a coherent policy system is that high achievement must count for something: grade promotion, high school graduation, college entrance, getting a job. Policymakers can ratchet up such requirements and expand the investment of resources to help some students keep up or catch up. The practical limits of any system, including limits on how long a student can engage in studies over time, will still have to be dealt with.

What makes standards *national*? Through the Goals 2000: Educate America Act, the Congress has authorized the creation of a National Education Standards and Improvement Council to evaluate and recommend to the National Education Goals Panel the certification of the standards developed by associations worthy of being labeled "national standards." The council also is to certify, on a voluntary basis, state standards. At this point, it is difficult to see why any state or professional association would want or need such labeling. The NCTM certainly doesn't need such an endorsement for the mathematics standards. Through its extensive consensus process involving thousands of math teachers, teacher educators, and mathematicians, NCTM achieved widespread endorsement by the profession that makes them national standards and an object of respect and acceptance among state policymakers, administrators, and textbook and test publishers.

2. Idea Two: Align the whole system of education policy to focus on teaching and learning that meet high content and performance standards.

This objective is based on the premise that classrooms and schools operate within larger district and state systems and that the rules, regulations, requirements, roles, responsibilities, and resource decisions of those jurisdictions have a large effect on teaching and students' learning. As currently functioning, the formal policy system of schooling in the U.S. is part of the problem. When initially articulated by Marshall Smith and Jennifer O'Day, systemic reform included (1) the state adoption of curriculum-content and student-performance standards in core academic subjects and the alignment with the standards of state policies and programs, such as staff development for teachers, to ensure that teachers possess sufficient knowledge to teach according to the standards, (2) providing textbooks and other materials that accommodate the teaching of the adopted standards, and (3) providing a student assessment system that measures whether the standards are being achieved. Smith and O'Day also addressed the importance of teacher discretion over matters of the "how" of teaching (Smith and O'Day 1991). More recently, those authors and others continue to expand the definition of "the system" and to include analyses of the political feasibility of policy coherence, the role of local school districts, policy and resource adequacy, and equity matters that affect students' opportunity to learn and other related issues (Fuhrman 1993).

Articulated here is the case for defining the system and changing its parts to make it more coherent. One can have, however, an efficient top-down bureaucratic system with closely aligned parts that does a good job of implementing a bad model. Restructuring governance, therefore, is dealt with in this essay as the third and necessary component of standards-based reform.

Antecedents

Coupled with a growing dissatisfaction with the results of schooling has been a growing awareness that much of the existing policy framework does not focus on results, certainly not on high academic expectations for all students. There has been little policy coherence. State and regional school accreditations are based on resource and process standards, not student outcomes. Student standards for promotion and graduation primarily reflect time spent, not knowledge and skill acquired. Standardized tests used to assess student achievement are not aligned with the curriculum content the schools are teaching but are designed to assign students to performance levels. The oars in the boat of school policy aren't pulling in the same direction. Analyses of schooling in the United States revealed that the loosely coupled system of schooling consists of policies and programs that aren't coupled at all. At the school level, many policies are ignored, but even more are followed and the results are fragmented, directionless efforts.

Many recent efforts to improve the system, primarily through policy change, have been searches for magic bullets and have had few or modest benefits. Begin some students earlier (more preschool). Improve resources (higher teacher salaries, lower pupil-teacher ratios). Require more academic courses (algebra for everyone). Make 'em accountable (high school exit examinations, school report cards). To mix metaphors, some state reform efforts have "bundled" several of these bullets into one (albeit unconnected) package. One personal example: Louisiana's Children First Act of 1988, which included higher salaries, smaller classes, more preschool education, and a state-administered process to evaluate teachers for purposes of periodic certification renewal.

The merits of each of these efforts can be rationalized, and we found some evidence of benefits. Preschool education for disadvantaged children does make a difference as evidenced by evaluations of the longitudinal effects of several state preschool programs; math achievement on standardized tests rose when all high school students were required to complete an algebra course. The magnitude of the benefits, however, was small and hardly represented a surge toward the learning of challenging subject matter by all students.

In the 1980s, many education reformers turned away from tinkering with the system and pursued a belief that improvement had to be school based and would occur one school at a time. Henry Levin's Accelerated Schools program,

James Comer's work, and Ted Sizer's Coalition of Essential Schools demonstrated that school-based and -designed reform can work. Hundreds of schools implementing an "effective schools" model achieved improvements in students' learning. In recent years, however, there has emerged discouraging evidence that a school-based, one-at-a-time reform strategy that ignores the policy context will not result in lasting change. Improvements dissipated when the project money ran out or the principal was transferred or some activity brought on the ire of a group of parents.

One experience gives hope to a systemic effort. What did make a difference for large numbers of students throughout the system occurred during the "basic-skills movement" of the 1970s and 1980s. The reform effort that focused on the learning of disadvantaged students had some serious flaws. The disaggregation of curriculum content into numerous small bits did not constitute challenging subject matter. It is likely that more-advanced students suffered; they certainly didn't benefit.

In urban districts like Atlanta, Jacksonville, and Birmingham, however, successive classes of elementary and secondary kids from poor backgrounds learned to read, write, and compute numbers far better than previous classes. What was characteristic of the plans implemented in those districts was the coherence or alignment of curriculum content, large-scale teacher in-service education programs, and the selection or design of instructional materials and development of student assessments. Many of us in the urban school districts during those years used the term *systems theory* in discussing our plans.

Growing dissatisfaction among state education leaders with the use of norm-referenced tests to measure the outcomes of schooling has supported the move toward a more coherent policy. For example, the insistence by Department of Education Secretary Ted Bell in publishing a "wall chart" to compare states using SAT and ACT tests crafted as predictors of college success, not a measure of school learning, led an increasing number of state officials to support the development of state-administered, criterion-referenced tests that were aligned with state curriculum frameworks and did measure what schools set out to teach. At present, local school district leaders express the same kind of dissatisfaction with norm-referenced standardized tests.

Because of the requirement in Chapter 1 of the Elementary and Secondary Education Act of 1985 that the progress of students receiving special assistance be assessed on an annual basis and because of the inexpensive availability of standardized tests, the use of such tests mushroomed during the 1970s and 1980s. In many districts, it became practical to spend a little additional local money and administer the tests to all students; some states financially support the testing of all students with standardized tests. As long as the results of such testing went only into annual Title I reports to Washington and to individual parents, little concern was expressed. When newspapers demanded and published standardized-test data comparing schools and school districts, the inappropriateness of such tests for measuring school outcomes became important to school administrators. Thus was laid the foundation for tests that were "authentic" reflections of what the schools were trying to teach.

More and more, local and state administrators were saying to each other, "If it is going to be done, let's be sure it is done right." Many school administrators were not reluctant participants in the movement to make the results of schooling public; they endorsed public accountability as an important and necessary feature of school improvement. They, too, wanted tests that measured the objectives of school programs.

Recognition grew in the 1980s that the content of the textbooks controlled instructional objectives in most classrooms and that, for the most part, textbook content was inappropriate (Cody 1991).

Issues

If the policy system is to be made coherent, what are the major parts of the system? The initial identifications are easy. One, determine the content and performance standards for each subject. Two, revise teacher certification and certificate renewal to insure that teachers gain and renew their mastery of what they need to teach. Three, support extensive staff development programs for teachers. Four, if textbooks are to continue to direct teaching, be sure that the content of the books and the instructional activities they include are appropriate for the standards. As a better alternative that was implemented during the "basic skills" reforms, support the engagement of teachers in crafting instructional guides based on the standards. Such handbooks thus serve teachers as a source of ideas for planning day-to-day classroom activities. Like other resources, textbooks become just another source of material to be used or not, as the individual teacher decides. Five, support the selection or development of student assessments that are aligned with the standards. Six, adopt processes that reward students and schools for success, and provide adequate assistance and appropriate consequences for lack of it.

The task gets a little more complicated when the logic that calls for policy coherence within the school system also suggests that schools are a part of other systems and that student success depends to some extent on child health and nutrition and the constructive handling of social and family problems. The coordination of community services on a voluntary basis among education and health and social service policy leaders is proposed by some. The National Science Foundation guidelines for urban systemic initiative proposals requires such coordination (National Science Foundation 1994).

State legislative mandates for agencies to work together may help. However, as a rise in the incidence of problems has coincided with the decrease in public funds to solve or prevent such problems, it may well be that the major need is not coordination but more dollars. Nevertheless, a coherent state policy framework for the education of children will likely warrant congruence with the policy frameworks that guide other social services.

There is yet another context in which schools and school systems operate. It consists of the "consumers" of those who complete the K–12 program: higher education and employers. The expectations of neither are well connected to schools. State boards of education can typically influence the nature of teacher education programs through their authority to approve such programs for teacher certification purposes. University admission standards, however, are another matter and are beyond the reach of elementary and secondary education policymakers. Yet they have considerable influence over what students choose to study in high schools and probably influence how hard they study. Who creates admission requirements varies from state to state; typically, state legislatures do not. Here, again, the articulation of policies by different policymaking bodies is warranted.

Can policies be constructed that induce employers in the United States to pay attention to the achievement record of high school graduates like employers do in Japan? At this time, very few do so. Recently, I asked the personnel officer of a major firm that employs many young people right out of high school why he did not use transcript information in his screening process; he replied that getting transcripts from high schools took too long. I suspect the problem is no more complicated than that. Nevertheless, a coherent policy framework might well include the creation of a new relationship between schools and employers that gives more value to grades as well as to graduation.

Finally, there is some reason to believe that the civic infrastructure of a community makes a difference in the quality of the schools and their success with students. Former Mississippi Governor William Winter has often described two communities in his state. One has the reputation of an outstanding school system and has some of the highest student achievement scores in the state; the other has some of the lowest. The communities are almost identical in size, racial diversity, per capita income, and percent of families in poverty. They have similar types of industries and employment patterns. In his observation, the only difference is in the number and variety of civic organizations that do good work and pay attention to the quality of life in the community and to the actions of elected and appointed officials.

The Kettering Foundation has an initiative to explore the relationship between the strength of a community's civic infrastructure and the quality of schools (Mathews and McAfee 1992). Clearly, there are things that policy leaders can do to engage members of the public in school affairs. It may well be, however, that public policies that support the creation and nurturing of a stable infrastructure of civic organizations for the benefit of the overall community will, in the long run, be more important than ad hoc involvement in occasional school matters. In order to flourish, schools may need to have a strong sense of community around them.

3. Idea Three: Restructure authority and responsibility in a way that delegates to teachers the discretion necessary for them to accomplish high standards.

This objective comes from the belief that for teachers to be effective and accomplish high standards with all students, they will need latitude and discretion over some matters that are now prescribed by state and local policies, programs, and resource decisions in the name of good public policy or good management.

Public policy must continue to address content and student performance standards, measures of outcomes for individual students and schools, standards on the "inputs" of schools that assure the opportunity to learn, the adequacy and the equitable distribution of resources, assurances of fairness and due process, and the authority and responsibility of staff at the central office and local school level.

In turn, local school staff, in exchange for being held accountable for achieving high standards with all students, should gain discretion over school schedules, definition of roles, organization of staff and scheduling of students' time, methods of teaching, and their own professional development.

Antecedents

The 1970s and 1980s witnessed a dramatic growth in the number of state laws and state and local policies that sought to make schools better. The public was becoming increasingly unhappy with the achievement of schools; chafing under the growing pressure, policymakers did what policymakers usually do—make more policy. Policy became more and more prescriptive of how schools should operate and how teachers should teach. Bureaucracies grew to be sure that the increasing requirements for teaching were met. It was a period during which "engineering" the process of instruction was carried to extremes.

During the same period, laws and policies were passed designed to hold teachers and schools accountable for results. In response, a small but growing number of school administrators and leaders of teachers organizations began to suspect that prescribing process and also holding teachers accountable for results was not only unfair, it was unproductive. The very nature of pedagogy and the differences in students require that teachers apply different methods with not only skill but also flexibility and discretion. Understanding this, Dade County, Fla., and Rochester, N.Y., administrators and teachers crafted new charters: Professional discretion and autonomy of teachers at the school site were granted in exchange for being held accountable for results with students.

In 1986, a task force of the Carnegie Forum on Education and the Economy published its report, *A Nation Prepared: Teachers for the Twenty-first Century.* The report presented a strong case for a new charter between policymakers and the teachers in the schools they oversee:

> Teachers should be provided with the discretion and autonomy that are the hallmarks of professional work. State and local governments should set clear goals for schools and greatly reduce bureaucratic regulation of school processes. Teachers should participate in the setting of goals for their school and be accountable for achieving agreed upon standards.

The growing interest in a new charter was influenced by the literature on schooling in America that used a factory metaphor. It described education's organization and structure as shaped by the industrial production model developed by Frederick Taylor in the early 1900s and made the case that the model was hardly appropriate to the professional process of teaching. The new role for teachers was also influenced by the growing conviction in private industry that the Taylor model of expert engineers designing the tasks to be precisely followed by workers, with quality control experts supervising the process to ensure things were done right, wasn't working well any more. Japanese automakers were building better cars than Americans, following a new industrial production model that vested quality control and the refining of the work task in the hands of the workers. The ideas behind total quality management in Japan were those of W. Edwards Deming, an American.

In addition, there is a logic inherent in the belief that student achievement is a function of expectations, opportunity, time, and pedagogy that supports the delegation of authority over some matters to the faculty of individual schools. Opportunity to learn, educational objectives, and equity issues might best continue to be addressed by the political body closely aligned with the public and in charge of overall resource allocation. Time and pedagogy, however, need to be varied and based on the needs of individual students. Each school—even each class and each student—will have different time and pedagogical needs. Those matters cannot reasonably be subject to uniform prescriptions from a central policy board. The logical extension is that the best policy is to leave the time schedule, staff and student organization, choice of teaching materials, and major as well as minor resource-allocation decisions to the school staff, where the variations needed are best understood.

Issues

What should be delegated to the professional staff of the school? Control over the school schedule and the allocation of students' time to various learning activities would seem to be the most important, and yet policies that regulate "minutes per week" for specified subjects for elementary students and time requirements for high school courses are almost universal. Authority to focus the full program of some students on a few core subjects seems necessary, as well as the resources to extend the time for those students to after-school, Saturday, and summer study.

The procedure for filling teaching vacancies should include approval by the school's faculty. If the faculty, as a group, is to be responsible, then that faculty should have a role in deciding who joins the group. Budget authority at the school level can be as minor an issue as deciding what instructional materials to purchase or as major as deciding how much of a school's budget to invest in personnel and how much in technology or other expenditures or whether to have twelve teachers and six assistant teachers rather than fifteen teachers. If and when a consensus develops over what should be delegated to the school level, it will likely evolve through experimentation and evaluation in the political arena as well as in the arena where evidence from research is honored.

Why would school boards be willing to delegate to teachers the authority for decisions they are now making? Some boards will be persuaded that school results will be higher with such delegation. More acceptance, however, will likely occur when as much attention is paid to the scope of the important decisions that the boards need to retain as to the decisions that need to be delegated to teachers. Approval of educational objectives, requiring evidence of results, providing adequate resources in an equitable manner, assuring due process and fair treatment, and other functions will continue to define decisions that should be made by representatives of the public.

Should the authority delegated to teachers at individual schools be, instead, the joint responsibility of teachers, parents, and citizens in the local school community? That model is being implemented in Chicago, Kentucky, and several other jurisdictions and is proposed for Alabama. Such a model seems a curious combination of the case for professional autonomy and a separate case for moving the governance of schools closer to the people they serve. Certainly public dissatisfaction with school system boards is at a high level; a not surprising response of state legislatures is to remove some board authority and give it to elected school councils. The authority proposed for school councils in Alabama, however, would provide teacher members equal voice with parents in curricular matters and parents equal voice with teachers in the scheduling of time and the in-service program of teachers.

Time and effort will reveal whether such councils will make things better.

CONCLUSION

Can a coherent educational policy based on the teaching and learning of challenging subject matter according to high standards of performance be created in the states and districts? The current policy structure seems to be more a function of diverse special interests having their will through the political process than a function of any rational plan for high achievement of all students. Nevertheless, the public demand for schools to accomplish more is stronger than it has been for many decades—strong enough, for example, to persuade a growing number of political leaders to search for a plan for reforms that will make a difference. It is significant that most state governors elected in the past ten years have made the improvement of elementary and secondary schools a major campaign and administration agenda. This was not so in previous decades. Education improvement has become a top agenda item for liberal, moderate, and conservative governors.

In closing this essay, I offer two observations on conditions that should prevail for standards-based systemic reform to be successful. First, it seems desirable, perhaps even necessary, for each state and school district pursuing standards-based reform to create a broad coalition of informal and formal business and civic leaders as well as educators and elected officials to help shape a reform agenda. Such a coalition should be institutionalized to provide continuity beyond the tenure of any one elected or appointed official. Current and recent activities in several states where the leadership is pursuing standards-based reform have revealed that there will be controversy.

The special interests and single-agenda advocates do not go away. In addition, the time required can easily be more than a particular elected or appointed official has to keep the central ideas alive. A coalition can help official leaders exercise moral courage in time of controversy and provide continuity past terms of office. A coalition of state legislators and leaders from business and industry has been central to keeping reform efforts in South Carolina on course over the terms of two governors. A similar coalition of political, business, and civic leaders has provided some stability in Kentucky. At the local level, the prevalence of factionalized school boards and the short tenure of superintendents call for a broad coalition of community leaders to keep reform at the local level on a standards-based track.

Second, a tendency to avoid dealing with some aspects of standards-based systemic reform because they will be problematic must be overcome. School boards may not want to delegate to teachers some matters the boards now handle, but the results of schooling will not increase much without that delegation. Reform efforts that address only on one or two parts of the system in the belief that everything else will fall in line will not likely accomplish much. There are no magic bullets. In my experience, the term *stasis* appropriately applies not only to the physical universe but also to social organizations like school systems. One dictionary defines *stasis* as "a condition of balance among various forces." The various parts of the system interact and are interdependent.

In a social system, any attempt to change one part will cause the system to exert energy to force compliance with the status quo by blocking, isolating, or expelling the source of change. Such is the dark side of bureaucracies. Such is the reality of political "agreements" that serve only special interests. Successful efforts at improving the education of *all* children, therefore, will address all three ideas in the paradigm: high content and performance standards, a coherent policy structure that focuses on the achievement of those standards, and a restructuring of authority that not only honors the public's right to govern a public service but also respects the need for teachers to have discretion over how to meet the public's expectations.

. . . .

REFERENCES

Bereiter, Carl., and Marlene Scardamalia. "Cognitive Coping Strategies and the Problem of Inert Knowledge." In *Thinking and Learning Skills: Research and Open Questions*, vol. 2, edited by Susan F. Chipman, Judith W. Segal, and Robert Glaser. Hillsdale, N.J.: Lawrence Erlbaum Associates, 1985.

Business Roundtable. *The Essential Components of a Successful Education System: Putting Policy into Practice.* New York: The Roundtable, n.d.

Carnegie Forum on Education and the Economy: Task Force on Teaching as a Profession. *A Nation Prepared: Teachers for the Twenty-first Century.* New York: Carnegie Corp., 1986.

Cody, Caroline B. "The Politics of Textbook Publishing, Adoption, and Use." In *Textbooks and Schooling in the United States*, edited by Arthur Woodward and David L. Elliott. Chicago: National Society for the Study of Education, 1991.

Education Commission of the States. *Introduction to Systemic Education Reform.* Denver: The Commission, 1992.

Elam, Stanley M., and Alex M. Gallup. "The Twenty-third Annual Gallup Poll of the Public's Attitudes toward the Public Schools." *Phi Delta Kappan* 71 (1991): 46.

Fuhrman, Susan H., ed. *Designing Coherent Education Policy.* San Francisco: Jossey-Bass, 1993.

LaPointe, A. E., J. M. Askew, and N. S. Mead. *Learning Science.* Princeton, N.J.: Educational Testing Service, Center for the Assessment of Educational Progress, 1992.

Mathews, David, and Noel McAfee. *An Alternative for Community Development and Education Reform.* Dayton, Ohio: The Kettering Foundation, 1992.

National Center on Education and the Economy. *America's Choice: High Skills or Low Wages.* Rochester, N.Y.: The Center, 1990.

National Commission on Excellence in Education. *A Nation at Risk.* Washington, D.C.: U.S. Government Printing Office, 1983.

National Council of Teachers of Mathematics. *Curriculum and Evaluation Standards for School Mathematics.* Reston, Va.: The Council, 1989.

National Council on Education Standards and Testing. *Raising Standards for American Education.* Washington, D.C.: U.S. Government Printing Office, 1992.

National Education Goals Panel. *National Education Goals Report: Building a Nation of Learners.* Washington, D.C.: U.S. Government Printing Office, 1991.

National Governors' Association. *From Rhetoric to Action: State Progress in Restructuring the Education System.* Washington, D.C.: The Association, 1991.

Smith, M. S. and J. A. O'Day. "Systemic School Reform." In *The Politics of Curriculum and Testing,* edited by Susan Furhman and Betty Malen, pp. 233–67. Bristol, Pa.: Falmer Press, 1991.

Task Force on Teaching as a Profession. *A Nation Prepared: Teachers for the Twenty-first Century.* Hyattsville, Md.: Carnegie Forum on Education and the Economy, 1986.

Westbury, Ian. "Comparing American and Japanese Achievement: Is the United States Really a Low Achiever?" *Educational Researcher* 21 (June–July 1992): 18–24.

Challenges for Education Policy at the Turn of the Century

Susan H. Fuhrman

IN THE mid-1990s, all levels of government are simultaneously engaged in a major reform movement. The goals of this movement are dramatic improvements in teaching and learning. The policy approach it features is standards-based reform, or systemic reform. While definitions of systemic reform vary (e.g., see O'Day [1995]), for many it means the establishment of challenging standards for student learning in key subject areas; coordinating central policy instruments, such as assessment, teacher certification, and teacher professional development, in support of the standards; and restructured governance to permit schools sufficient flexibility to meet high standards.

Systemic reform is most visible as a state-level movement. In the mid-1980s, California and other leading states began to develop expectations for what students should know and be able to do. By 1993, as many as forty-five states claimed to be in the standards and assessment business (Pechman and Laguarda 1993). Twenty-six of them have been receiving National Science Foundation (NSF) support under the Statewide Systemic Initiatives Program. The Council of Chief State Officers and the Education Commission of the States took support for systemic reform as their key theme for working with states; the National Governors' Association, the Business Roundtable, and other leadership associations have also promoted the concept.

Systemic reform is also a national-level movement. NCTM pioneered the notion of establishing challenging expectations for all students. Its 1989 *Curriculum and Evaluation Standards for School Mathematics* has served as a model for other disciplinary groups that are now trying to reach a consensus

Susan Fuhrman *is director of the Consortium for Policy Research in Education. She has written numerous articles, research reports, and monographs on education policy and finance and is the editor of two books on systemic reform:* Designing Coherent Education Policy: Improving the System *and* The Governance of Curriculum.

about standards for student learning. The Clinton administration has made systemic reform the organizing principle for its education policy efforts. Goals 2000, signed into law in March 1994, sets the framework. It will support the development and certification of voluntary national academic standards in core subject areas such as science, math, history, English, geography, foreign languages, and the arts. Funds are provided for the development of assessments to measure achievement of the standards.

Goals 2000 also provides grants for state and local reforms, such as the development of standards, assessments that measure attainment of the standards, related changes in teacher certification, enhanced professional development, improvements in technology, and changes in governance and accountability. The reauthorized Elementary and Secondary Education Act also calls for standards and the integration of other policies with the standards. The standards used for Title I, Indian Education, and other federal programs would be those set by the states to govern their own programs. At each level of government, key policies—assessment, teacher professional development, and so on—would be linked to the standards (Fuhrman 1994a).

At the same time, many local districts are going about their own standard-setting activities and developing related reforms (General Accounting Office 1993). Local systemic reform efforts are given a big boost by the National Science Foundation's new Urban and Rural Systemic Initiatives Programs and the Goals 2000 local grants.

This policy movement provides many challenges. It is very complicated and difficult, since it involves redesigning many aspects of policy at the same time and doing so in a way that integrates previously discrete, and sometimes contradictory, policy instruments. Other chapters in this volume will concern important aspects of improving support for schooling, such as building public engagement and enhancing professional development. In this chapter, I focus on policy-design challenges that must be confronted both in the short term, to get reforms off the ground, and in the longer term, to institutionalize reform efforts. I group the challenges into the following headings: (1) designing new processes and structures, (2) incentives, (3) accountability, (4) school finance, and (5) opportunity to learn.

DESIGNING NEW PROCESSES AND STRUCTURES

As important as specific policy instruments—such as content and performance standards and authentic assessments—are to this reform movement, the creation of legitimate processes to develop and implement such key reform elements is equally critical. Reform components require extensive processes of consultation, comment, and review. For example, in establishing commissions to develop content and performance standards, states are trying to incorporate both widespread public input and professional expertise, represent diverse populations and perspectives, manage subdisciplinary schisms within certain subject areas, and, in many cases, seek links across subject areas. Not only must such processes be carefully designed to accomplish initial standard-setting and related activities, they must also form the starting point for dynamic efforts that refine standards over time to incorporate new knowledge (Consortium for Policy Research in Education 1993; Massell 1994; Massell and Kirst 1994).

Process is particularly critical to these reforms for at least five reasons. First, to be credible expressions of societal expectations for student learning, standards and the other policies that support them must represent public input on a significant scale. There is evidence that many parents and citizens do not understand or believe in the principle undergirding standards-based reform, that all children can learn to much higher levels (e.g., Farkas 1993). Involving parents and citizens in reform design is only a small part of the public education effort that must take place to overcome confusion and generate public support, but it is a very important part.

Second, to provide leadership for improvements in teaching and learning, policies must also represent expert professional judgment. For example, the California subject-matter frameworks have great legitimacy and respect in the field because of the professional consensus they reflect, and evidence of their use is growing (Massell 1994; Cohen 1990; Peterson 1990). Policy design efforts must emphasize professional leadership, such as that provided by NCTM in its groundbreaking effort to develop consensus around student expectations. Teacher involvement in design is also necessary to keep the reforms from being top-down directives that constrict rather than enable school practice. To develop policies that support ambitious instruction, many states are enlisting teachers to serve on standards or curriculum framework commissions, develop assessment tasks, score assessments, and undertake other central reform responsibilities (Massell 1994; Koretz et al. 1994).

A third reason for attention to process and structures for policy development is that the proposed reforms involve links across areas of policy traditionally governed by separate bodies. For example, to assure that teachers are prepared to teach what the new standards propose for student learning, state education agencies, professional standards boards that control teacher licensure, higher education governing bodies, and higher education institutions need to work together. Some states may find that new cross-agency structures or coordinating bodies are necessary to make such critical connections.

In several states participating in the NSF Statewide Systemic Initiatives Program, for instance, new quasi-governmental or nongovernmental organizations are playing key leadership roles. Some of these, like the Kentucky Science and Technology Council and the Montana Council of Teachers of Mathematics, are not new organizations, but they are taking on new responsibilities. Others, such as the Council for the Advancement of Mathematics and Science Education in New Mexico and the Connecticut Academy for Education in Science, Mathematics, and Technology are new broad-based collaboratives created to promote and extend reform. Such organizations are of particular interest because of their unique relationship to the formal decision-making hierarchy in education and because participants say that they broaden support for reform, particularly among teachers and business leaders (Shields, Corcoran, and Zucker 1993).

Fourth, if reforms are to last over time, ways must be found to sustain momentum across changes in political leadership. But standards and related policies must also be revised over time to accommodate new knowledge and improvements in understanding about student learning and instruction. The reforms must be both stable and dynamic. The current political process—driven by short electoral cycles, prone to layering new policies on top of old instead of revisiting and improving existing approaches, and fragmented structurally both within and across layers of government—does not naturally support such a balance.

Some states are experimenting with new entities intended to maintain reform momentum over time. For example, Kentucky's reform legislation vests oversight responsibility in a new Office of Accountability. Its placement in the Legislative Research Commission might serve to cement legislative commitment to the reform act; legislators become the keepers of the act, charged with monitoring and fine-tuning it, and may be less likely than otherwise to forge off in new directions. Kentucky also has a ninety-five-member citizen's body, the Prichard Commission, that reports on reform progress and engages in public education. Like South Carolina's Business-Education Subcommittee, which has existed since that state's reforms of the early 1980s, it unites political and citizen leaders behind a vision of reform that is intended to bridge changes in leadership and political fortunes (Adams 1993; Fuhrman 1993).

Fifth, as must already be evident as this essay proceeds, reform is an immensely complicated activity and capacity is sorely lacking. Many states and localities are finding the work of designing multiple policies at the same time in a coordinated fashion immensely difficult, especially since recession-related cuts in agency staff have eliminated many of the curriculum specialists and policy generalists best suited to such activities.

One particularly vexing issue, for example, is how to sequence reforms. A number of states are running into problems—using old tests in the midst of new content expectations or new tests before teachers or students are adequately prepared. To build the capacity to address such questions and to accomplish the work of reform, policymakers are exploring new processes, such as networks of teachers and university faculty who conduct some of the development tasks associated with reform. In Vermont, for example, state agency leadership has created outside capacity by generating networks of professionals working on assessment development and scoring. In Michigan, strong universities and professional associations have provided leadership.

New commissions, collaborations, networks, and the like are ways to accomplish the work of reform, include wider groups of reform participants in design and development, and communicate across agencies and electoral cycles. At the same time, however, coordinating the variety of proliferating commissions is no easy task. It will be important to watch these entities over time to see if they can bridge fragmentation without adding yet additional layers to an already complex governmental structure (e.g., Morone 1990).

INCENTIVES

Many reformers believe that well-conceived and publicly supported systems of standards and related reforms speak to some of the most fundamental, intrinsic motivations of everyone involved in the educational enterprise. Standards-based reforms address a desire to promote learning and to see evidence of progress. Because common expectations for student learning should represent societal agreement on important schooling outcomes, standards could bring more meaning and reinforcement to classroom work. In addition, assessments linked to the standards could model expected performance for the students, supply welcome feedback to teachers, and provide a means for legitimate accountability based on school and system improvement toward clear outcome expectations. Criteria for success would be apparent and collectively endorsed; success would therefore be more easily identified and celebrated. Over time, as more and more citizens share the standards and as what it takes to make progress toward them becomes clearer, the motivation to achieve them would grow. (See Firestone and Pennell [1993] for a review of the literature on teacher incentives.)

However, the reforms cannot begin and end with standards and assessment, as much current policy rhetoric suggests. If standards and assessment are all there is to it, then it's hard to imagine how teaching and learning will actually improve. It seems—and in fact is—nonsensical to imagine that setting higher goals and measuring their attainment will, in and of themselves, lead to improved teaching and learning. Policymakers need strategies for moving from standards and assessment to improved classroom practice, or more accurately, for designing the array of policy approaches that need to complement standards and assessment if the policy system is to provide sufficient support for instructional improvement.

Much more could be done to enhance motivation throughout the system. Some suggest that student employment opportunities and college admission decisions could be linked more closely to school performance. To encourage teachers to take more responsibility for student learning, we could bolster the professional nature of teaching. Approaches might include workplace redesigns that speak to teachers' needs for autonomy and participation; reforms in preservice education and in professional development to give teachers the opportunities and skills to function as collaborative, continuous learners; and professional self-regulatory mechanisms for teacher and school evaluation. Other possibilities include recognition for teaching excellence through mechanisms such as the National Board for Professional Teaching Standards and compensation systems designed to reward knowledge and skill acquisition rather than random credit accumulation (Odden and Conley 1991).

For schools and systems, incentive design should extend beyond monetary rewards for improvement, although those might be part of the package. Currently six states (Georgia, Indiana, Kentucky, North Carolina, South Carolina, and Texas) provide monetary rewards for improved school performance. The number of states providing such awards has never risen above ten, suggesting the political and technical difficulties of instituting such rewards.

Another approach to incentives focuses on removing disincentives to change, like finance formulas, program regulations, or union contract provisions that lock schools into certain staffing ratios and organizational patterns. For example, states could revise formulas based on classroom or teacher units that pose barriers to more flexible class sizes and scheduling. Accounting systems that clarify and permit public understanding of school resource use might be considered, as might governance changes such as choice and charter schools that alter the relationships between providers and consumers.

One important reason to think about incentives in a sophisticated fashion relates to a fundamental purpose of current reforms: to make systemwide the kind of improvements that past reform movements have brought to no more than 10 to 15 percent of schools at a time (Elmore 1991). In many states, new standards-based systemic reform efforts can be found in geographic pockets, such as pilot sites, or at certain grade levels, or in certain subject areas. In a number of states, technical and policy efforts appear to run on separate tracks. Some excellent pilots, in professional development, for example, exist in states that have not developed policies to support such efforts on a broader scale, whereas other states have policies in place without the good technical progress worth generalizing.

A key challenge in generalizing these approaches is figuring out what motivates the range of teachers, schools, and districts. Policymakers typically treat the problem of scaling up from a few innovative sites as an educative endeavor and see their role as providing other schools time and opportunity to learn from model sites. But not everyone responds to the enticement of greater knowledge; if the goal is important enough, other options, eventually including mandates and sanctions, might be warranted (Elmore 1994).

In summary, policy encouragement for maximizing achievement on standards involves complex policy-design challenges. The array of incentives in current policy systems should be analyzed for their fit with reform goals and their power. Disincentives should be removed and new policies, governance approaches, and financing schemes designed to address the range of possible responses to standards and assessments.

ACCOUNTABILITY

Another area where new thinking about policy is needed is accountability. More and more states are reporting student outcomes (aggregated to the school or district level) publicly, including performance measures along with compliance measures as criteria in school or district accreditation/certification reviews, and attaching rewards and sanctions to school performance. Whatever their specific approaches to rewards and sanctions or to accreditation review, all states are facing a number of difficult issues in moving toward performance-based accountability.

One issue is the lack of good outcome measures. Since new assessments are still under development in many places, there are not always good measures to rely on. So, for example, policymakers may wish to eliminate their focus on compliance with requirements that schools offer so much science or social studies, but they are reluctant to do so if there are no measures of science and social studies outcomes to substitute. Relatedly, it is hard to rely on brand new assessments—perhaps with some questions about reliability and validity not yet resolved—to the extent necessary for accountability (McDonnell 1994). Unfortunately, the hesitancy to attach consequences to new assessments is not as widespread as educators might like. There is danger that the whole reform movement will falter on the rush to stakes because the public and educators will resent the fact that schools and students are not given sufficient opportunity to learn the new standards.

Another obstacle to the redesign of accountability is policymakers' worry about habitually noncompliant districts, those frequently termed "bad apples." They are very reluctant to move away from enforcement of regulation through traditional compliance reviews for places like Jersey City or Newark or for districts in eastern Kentucky where patronage, nepotism, and generalized corruption are widespread (Dolan 1992; Braun 1993; Fry, Fuhrman, and Elmore 1992). The question of what to do with such districts is one of the most perplexing and troublesome state policy problems. Typically, policymakers make regulations for the whole state with these troubled places in mind; this mindset goes a long way toward explaining why education codes are lengthy and frequently restrictive of practice.

Closely related is the question of what to do about failing schools, whether or not they are located in corrupt districts. New approaches to problem schools, such as closing and reconstituting schools and offering parents other choices, merit careful study. Finally, policymakers are struggling with how to foster an ethic of continuous improvement in the context of accountability. How do we move away from binary accountability (a school is accredited or not) to a system where schools set goals that improve over time and figure out how to get there as organizations?

Clearly, incentives and accountability are very closely related. Any accountability system incorporates numerous assumptions about incentives. Newer approaches assume that public reporting of school-level test scores, monetary rewards, and fear of punitive sanctions are strong motivators for maximizing student achievement; yet these assumptions are relatively unresearched. Are there important differences in incentives created when stakes are attached to basic-skills tests (where it's relatively easy to figure out how "teaching to the test" may improve scores) as opposed to newer performance-based assessments (where improving performance may require knowledge and skills many teachers lack)? How can professionally run examinations of teaching and learning, as New York and other states are developing to shed light on the complexities of practice, balance the conflicting demands of professional growth and accountability? These are difficult questions, requiring study and reflection as well as imaginative policy design.

SCHOOL FINANCE

An important need is to create a funding system that supports the current reforms (Furhrman 1994b). There is no dollar figure that represents "adequate" support for the kind of reforms in teaching and learning policymakers and educators are seeking. We do not yet know the extent or cost of the changes in policy and practice that will be necessary to make the reforms feasible. But at this point it is possible to identify many reform components that will require serious investment—from new funds, from reallocation of existing resources, and from the coordination of previously discrete pots of money from various levels of government and private sources. These include the development of high-quality curriculum frameworks, assessments, materials, and professional development approaches; making time available to assure that both teachers and students have adequate opportunity to learn; and providing extra support for the varied educational and other social service needs of disadvantaged children.

Beyond the amount of money, the way that money is allocated also has implications for reform. Right now, finance and policy tend to run on separate parallel tracks, with finance a function of formula-driven allocations that are not tied to policy goals.

Courts are trying to change that relationship by increasingly focusing on the outcomes and programs in equity-related court cases. Some cases explicitly use outcomes as evidence of the effects of unconstitutional finance systems; others refer primarily to programs that are expected to influence outcomes (Clune 1993).

Rose v. Council for Better Education (1990) in Kentucky is probably the most well known case to rely on evidence about outcomes. The court considered the state's ranking both in the South and nationally in deciding the system was inadequate. It also compared test scores among districts of varying wealth and cited expert testimony linking achievement and resources. In 1993, Alabama's court concluded that lack of student success in graduating and in postsecondary endeavors reflected badly on the system, which was declared inadequate as a result. In Montana, Texas, and New Jersey, the focus was on wealth-related disparities among districts in programs and offerings, which were expected to relate to student performance, not on revenues or expenditures in isolation. This adequacy focus is quite a change from the first round of cases in the 1970s and early 1980s. Earlier school finance cases centered on equitable funding and invalidated systems based on the degree of wealth neutrality, not on the degree of preparation afforded students.

Policymakers are just beginning to think about how to achieve the link between dollars and policy goals directly through finance formulas. For example, at least two states are using formula aid to support professional development, a particularly important component of current reforms (Little 1993). As teachers grapple with the content and more-complex pedagogy implied by challenging student standards, the need for investment in professional development becomes clear. Many teachers lack sufficient subject-matter preparation; most have little familiarity with teaching approaches that ask them to facilitate learning rather than transfer content. But staff development was frequently seen as an extra not meriting support in tight fiscal circumstances. As a separate budget item, it was very vulnerable to cuts. Now Minnesota and Missouri have directed that a portion of school funding be set aside for this purpose, indicating that political realities are shifting and that states are becoming willing to push policy goals directly through the general state aid formula.

Some reforming states are starting to examine the role of the school finance formula in constraining school behavior. For example, some policy-makers in Delaware, which has a formula based on shares of teacher salaries and other staffing ratios, are planning to examine the formula in light of its New Directions reforms, which promote ambitious outcome expectations and stress streamlining of impediments to practice. It is likely that a number of states will, at some point in the future, consider whether formulas need to be redesigned so that they do not restrict school behavior.

To some extent, discussion about revising formulas to create incentives for policy goals is still talk, largely confined to policy experts and analysts (e.g., Odden 1993, 1994). So while academics might suggest that states funnel money directly to schools, not districts, since schools are the unit responsible for achievement, no state has yet taken such a step. But the developments just cited indicate the beginnings of movement away from the entitlement conception of school finance formulas toward the idea of formulas that build in provisions to promote policy goals.

OPPORTUNITY TO LEARN

The goal of high standards for all students carries with it a moral obligation to provide opportunities for all students to reach the standards (O'Day and Smith 1993). Providing equitable opportunity in the context of systemic reform involves its own set of policy and political challenges, above and beyond continuing, very serious finance and resource inequities that exist in most states and within many districts as well.

One challenge concerns the definition of "high standards for all students." In most states, there has been little discussion about what a good result would be from a distributional point of view. What weight is accorded to the goal of progress for all as opposed to closing achievement gaps between minorities and white students? Would we settle for the same distribution if everyone moved up to higher levels? If there is a minimal high standard for all (if we can juxtapose those words), what does the distribution above it look like? The policy activity around high standards has not yet incorporated frank talk about what that means for all students.

Related to this avoidance is the absence of serious reaction to achievement gaps in states with new interim assessments that incorporate much higher standards. Overall achievement on such new tests is generally quite low. (In Delaware in 1993 only 2 to 11 percent of the students in the grades assessed met or exceeded the standards for writing proficiency; 13 to 22 percent were at least proficient in reading; and 11 to 17 percent were proficient in mathematics.) The press and public have focused on how poorly everyone is doing, on how much more revealing the new tests are about actual levels of achievement, and on how the reforms offer hope for improving average performance. Even when scores are disaggregated, not much attention is paid to gaps, perhaps because everyone, including parents of traditionally high-scoring kids, is preoccupied by the overall low levels.

A FAIR CHANCE

Beyond talking about what good achievement for all students means, a major task for policy is seeing that everyone has a fair chance to get there. In congressional debates about Goals 2000 and the reauthorization of the Elementary and Secondary Education Act, there was considerable discussion about "opportunity to learn standards." Some, particularly some key House Democrats, are so concerned that setting high outcome standards could unfairly disadvantage students in less well off schools that they wanted to direct states to develop relatively specific opportunity-to-learn standards. They argued that no content and performance standards states develop as part of their reform strategies should be used for policy purposes or for making major decisions about student promotion, graduation, and retention unless opportunity-to-learn standards are in place. Others, including many Republicans, prefer to take a more cautious approach, focusing on research and development rather than immediate standard setting. They want states to address opportunity to learn but in a less specific and directive manner. The second position had obvious appeal to state policymakers who are worried about federal dictation of state policy.

What the congressional discussions did not sufficiently acknowledge was the fact that states have been trying to assure opportunity to learn throughout most of their history, primarily through finance equalization and minimum standards for service provision. The fact that they have not been very successful, as school finance and desegregation litigation illustrates, indicates that regulating or setting standards about services is not a very promising approach for assuring opportunity to learn. New regulations should not be our first choice of policy instrument for assuring equity. This is especially true because the focus of current reforms on performance carries with it the goal of removing, not adding to, regulation that constrains schools in reaching the standards (Elmore and Fuhrman forthcoming).

We need to operate with caution in this arena. The only thing we know with certainty about various solutions to the opportunity-to-learn problem is that traditional input-focused regulations have not worked. States should probably be looking for mechanisms that nudge state policies generally in the direction of greater reliance on performance controls rather than input controls while looking for more effective and less obtrusive ways to engage in input regulation. One possibility would be increasing reliance on professional standard setting and self-regulation as instruments of state policy. School quality reviews, such as those just getting under way in New York State, may develop standards of best practice that can be applied to judge opportunity (Elmore and Fuhrman forthcoming). Finally, indicators of opportunity to learn, such as measures of enacted curriculum and pedagogy, can help policymakers assess the extent to which all children are exposed to good practices and high-level content (Porter 1994).

Assuring that reforms are truly equitable will require both imaginative policy design and vigilance over time. As with incentives, accountability, and finance, we will need to monitor opportunity-to-learn strategies to assess the extent to which they support the reforms as they develop and evolve. All the more reason, then, to pay careful attention to the design of structures and processes—the first challenge identified in this paper—so that policy-makers have ways of tracking and fine-tuning reforms. As complex as systemic reform is in the short run, the challenges it poses will be with us for many years as the hard work of changing policy to support teaching and learning continues.

· · · · · · · · · · · · · · ·

NOTES

This article draws on research conducted by the Consortium for Policy Research in Education, a joint venture of Rutgers, The State University of New Jersey; the University of Wisconsin at Madison; Harvard University; Stanford University; and the University of Michigan, funded by the U.S. Department of Education's Office of Educational Research and Improvement (OERI-117G10007). The views expressed are those of the author and do not represent those of the consortium or its sponsors.

· · · · · · · · · · · · · · ·

REFERENCES

Adams, J. E. *The Prichard Committee for Academic Excellence: Credible Advocacy for Kentucky Schools*. New Brunswick, N.J.: Consortium for Policy Research in Education, 1993.

Braun, R. J. "Schools that Flunk." Star-Ledger Special Report series. *Newark (N.J.) Star-Ledger*, 24–31October 1993.

Clune, W. H. "The Shift from Equity to Adequacy in School Finance." *The World & I* 8 (9) (1993): 389–405.

Cohen, D. K. "Revolution in One Classroom: The Case of Mrs. Oublier." *Educational Evaluation and Policy Analysis* 12 (3) (1990): 327–45.

Consortium for Policy Research in Education. "The Process of Developing Standards for Student Learning." *CPRE Policy Briefs*. New Brunswick, N.J.: The Consortium, 1993.

Dolan, Margaret. *State Takeover of a Local School District in New Jersey: A Case Study*. New Brunswick, N.J.: Consortium for Policy Research in Education, 1992.

Elmore, R. F. " 'Getting to Scale' with Good Practice." Working paper for the Research Forum on Incentives in Education, 1994.

_____. "Innovations in Education." Paper presented at Innovation in the Public Sector Meeting, Duke University, Durham, N.C., May 1991.

Elmore, Richard F., and Susan H. Fuhrman. "Opportunity to Learn and the State Role in Education." In *Teachers College Record*, forthcoming.

Farkas, S. *Effective Public Engagement*. Paper prepared for the New Standards Project Conference by the Public Agenda Foundation, Rosemont, Ill., 1993.

Firestone, Willliam A., and James Pennell. "Teacher Commitment, Working Conditions, and Differential Incentives." *Review of Educational Research* 63 (4) (1993): 489–526.

Fry, Patricia, Susan H. Fuhrman, and Richard F. Elmore. *Schools for the Twenty-first Century Program in Washington State: A Case Study*. New Brunswick, N.J.: Consortium for Policy Research in Education, 1992.

Fuhrman, Susan H. "Clinton Education Policy and Intergovernmental Relations in the 1990s." *Publius* 24 (1994a): 83–97.

_____. *Education Policy Issues and Data Needs in School Finance*. Washington, D.C.: National Center on Education Statistics, 1994b.

_____. "The Politics of Coherence." In *Designing Coherent Policy: Improving the System*, edited by Susan H. Fuhrman, pp. 1–34. San Francisco: Jossey-Bass, 1993.

General Accounting Office. *Systemwide Education Reform*. Washington, D.C.: GAO, 1993.

Koretz, Daniel, Brian Stecher, Stephen Klein, and Daniel McCaffrey. *The Vermont Portfolio Assessment Program: Findings and Implications*. Washington, D.C.: RAND Institute on Education and Training, 1994.

Little, J. W. "Teachers' Professional Development in a Climate of Educational Reform." *Educational Evaluation and Policy Analysis* 15 (2) (1993): 129–51.

Massell, Diane. "Setting Standards in Mathematics and Social Studies." *Education and Urban Society* 26 (2) (1994): 118–40.

Massell, Diane, and Michael Kirst. "Determining National Content Standards: An Introduction." *Education and Urban Society* 26 (2) (1994): 107–17.

McDonnell, L. M. *Policymakers' Views of Student Assessment*. Washington, D.C.: RAND Institute on Education and Training, 1994.

Morone, James. *The Democratic Wish*. New York: Basic Books, 1990.

National Council of Teachers of Mathematics. *Curriculum and Evaluation Standards for School Mathematics*. Reston, Va.: The Council, 1989.

O'Day, J. A. "Systemic Reform and Goals 2000." In *National Issues in Education: Goals 2000 and School-to-Work*. Bloomington, Ind.: Phi Delta Kappa, 1995.

O'Day, J. A., and M. S. Smith. "Systemic Reform and Educational Opportunity." In *Designing Coherent Education: Improving the System*, edited by Susan H. Fuhrman. San Francisco: Jossey-Bass, 1993.

Odden, Allan. "Including School Finance in Systemic Reform Strategies: A Commentary." *CPRE Finance Briefs*. New Brunswick, N.J.: Consortium for Policy Research in Education, 1994.

____. *Redesigning School Finance in an Era of National Goals and Systemic Reform*. Washington, D.C.: National Alliance of Business and Education Commission of the States, 1993.

Odden, Allan, and Sharon Conley. "Restructuring Teacher Compensation Systems to Foster Collegiality and Help Accomplish National Education Goals." In *Rethinking School Finance: An Agenda for the 1990s*, edited by Allan Odden, pp. 41–96. San Francisco: Jossey-Bass, 1991.

Pechman, E. M., and Katrina Laguarda. *Status of New Curriculum Frameworks, Standards, Assessments, and Monitoring Systems*. Washington, D.C.: Policy Studies Associates, 1993.

Peterson, P. E. "Doing More in the Same Amount of Time: Cathy Swift." *Educational Evaluation and Policy Analysis* 12 (3) (1990): 261–80.

Porter, Andrew. "Defining and Measuring Opportunity to Learn." In *The Debate on Opportunity to Learn Standards: Supporting Works*. Washington, D.C.: National Governors' Association, 1994.

Shields, P. M., T. B. Corcoran, and A. A. Zucker. *The Study of NSF's Statewide Systemic Initatives (SSI) Program: First-Year Report*. Menlo Park, Calif.: SRI International, 1993.

The Business and Industry Perspective

David Kearns

I DIDN'T get interested in education out of some social or esoteric concern. I got interested as a businessman—from my perspective as CEO of Xerox. The bottom line is that I was worried that my company wouldn't have the workforce with the skills—the *technical* ability and the *adaptability*—to compete in the twenty-first century. We were losing our competitive edge as a nation.

The truth is that our schools are not adequately preparing our students to compete. Something's happened to American education in the past thirty years that has had a direct impact on the performance of American business and industry.

What's happened is that we've been standing still while the rest of the world has marched past us. Studies show that our children today are learning at a 1970 level. So it's not that our schools have gotten worse, it's just that they've stayed the same. The Summer 1994 report of the National Assessment of Educational Progress pointed to modest gains in student achievement in mathematics: 28 percent of nine-year-olds can now add, subtract, multiply, and divide whole numbers, up from 20 percent in 1978. That does not strike me as a record to be proud of yet. Yet more discouraging is the fact that only 7 percent of our seventeen-year-olds can solve multistep problems and use basic algebra—a number that has remained stagnant for over twenty years. So it's not that our schools are getting worse, or that our kids are slow—it's simply that the competition has improved dramatically and continues to do so at a geometric rate. I'm not pessimistic about our schools, but the point is they just aren't good enough.

David Kearns is chair of the New American Schools Development Corporation, a private, nonprofit, bipartisan organization, and is a senior fellow at the Harvard Graduate School of Education. He has served as deputy secretary of the U.S. Department of Education and was chair and CEO of Xerox Corporation.

The rest of the world hasn't been standing still. Twenty years ago, Korea was a third-world country. Today, it's our competitor. Twenty years ago it only exported plastic toys. Today, it competes head to head with us in production of cars, heavy industry, and telecommunications. And the competition is only going to get tougher.

As CEO of Xerox in the 1980s, I knew that American companies were falling behind in critical high-tech industries—the industries that would define the workplace and the workforce of the twenty-first century. In fact, I thought there was a good chance that Xerox would go out of business on my watch. It didn't take a rocket scientist to figure out why. It took us twice as long as the Japanese to develop our products and bring them to market. They were selling their copiers at what it cost us to build ours. And their quality and reliability improved every year while ours remained about the same. We took comfort in 7 percent improvements in productivity when in fact we needed 17 percent if we were to compete with the Japanese. Essentially we had been standing still for years while our competition was improving steadily. We needed to dramatically overhaul our company if we were going to stay in business.

These improvements could not be achieved by tinkering around with our business processes; they required wholesale change in the way we did business. And we needed a workforce that could lead that scope of change. It was at that point that I—and my company—made a commitment to education as the fundamental underpinning of our competitiveness and that of the nation. It was simply a matter of survival.

Since then, Xerox has gone through a wrenching process of change that has put the company back on top of the industry. We cut our unit manufacturing costs by 50 percent. We cut one whole year out of our development cycle. We invested heavily in benchmarking and employee training. And we restructured our business processes to empower our employees so as to keep pace with the competition.

If the scale of these improvements were "2X," what's needed now is "10X." Paul Allaire, my successor, now believes that if Xerox is going to continue to beat the competition, it will have to increase that scale of improvement tenfold. And as the demands of improvement and competition increase, so, too, will the demands on our business structure, processes, and employees.

XEROX IS NOT ALONE

What has happened at Xerox is happening throughout business and industry. The businesses of the future will have to be lean and mean—able to adapt to ever changing markets. There will be less hierarchy, less managerial review, a more horizontal structure. These businesses will have to be structured to take advantage of specific opportunities and to encourage innovation and speed. They will have to empower employees, giving them all the tools and information to get the job done fast to take advantage of increasingly smaller windows of opportunity. Big, monolithic, highly structured companies, with hundreds of thousands of employees, will be replaced by small, nimble, upstart businesses.

Likewise, these businesses will require a workforce that can work without such oversight, that can move quickly, that can meet each specific challenge. They will require a technical workforce with instant access to information and power to meet the challenge of the rapid marketplace. Old manual jobs will be replaced with a new era of technicians, keeping pace with technology.

Look what's happened to our automotive industry in the past twenty years. Yes, there are fewer assembly-line jobs, but those jobs require a much higher degree of skills. And the workplace of tomorrow will be increasingly demanding. There will be those who have the skills required to be a part of that workforce and, sadly, those who don't. For those who do, their quality of life will only improve, and for those who don't, they can only expect to stay about the same, if that.

The companies of the future will look a lot like the small, flexible companies of today—companies like SOFT-DESK in Henniker, New Hampshire. SOFT-DESK has 200 employees worldwide developing software programs for engineers and architects. They all interact via computer, even though employees may work together on any given day on the same project from Belgium; San Diego, California; Troy, New York; or Henniker. The company is structured to be adaptable—there are three levels; the CEO, three vice-presidents, and everyone else. They are all cross-trained and "multitasked," so they all play a role in development, marketing, or service. They invest heavily in training and utilize technology so their employees are "multipliers"—making up for their small size through technology and their abilities. And as the technology advances in the industries they serve, so do their customers' requirements, necessitating continued adaptability and training.

In 1970, Henniker was primarily a lumber community. Most of the skills its graduates needed for life were learned in shop class. That's not the case today; if you want a good job in Henniker, chances are you'll want to work for SOFT-DESK. You won't necessarily need a college degree, but you'll need to be computer literate. That's certainly the case throughout America in towns like Pittsburgh, where industry has changed dramatically and so, too, have the requirements of employment. Much of our steel production has been lost to overseas companies, and in its place is a new generation of small, high-tech industries. The workforce had to go back to school if it wanted to play a part. They did in Pittsburgh, and the city now is rated as having one of the highest qualities of life in America.

It all gets down to competitiveness. Those businesses that want market share are going to have to keep up with change and technology; those employees that want to be a part of that workplace had better have the skills to compete; those nations that want to lead the competition had better have the skilled workforce. The requirements will be technical ability, adaptability, and an ability to meet the standards of the marketplace. The return will be a higher quality of life, products, and services.

Our students—our future workforce—need to know what is required of them to take part in this workplace revolution. *Standards* for me means identifying what these requirements are—in other words, what it is that you need to know and do in order to compete. At Xerox, we benchmarked. We looked throughout business to see who was the best in a given discipline and then sought to do better ourselves. In education, we need to do the same sort of thing. We need to ask ourselves, "What is it that our children need to know and be able to do to be the best in the world?" In particular, what do they need to know to be a part of the *technical* workforce? Answering these questions and developing standards to meet them are critical steps in education reform.

STANDARDS SHOULD REFLECT THE REAL WORLD

These standards ought to reflect what is required in the real world. In West Bend, Wisconsin, the schools operate a youth apprenticeship program for high school juniors and seniors with local printing businesses. The students started working and learning how to operate the equipment, yet they were having trouble with the math skills required for the statistical process controls. The schools changed the math curriculum to include more statistics and more examples from the real world. The students' performance improved one whole letter grade on average. Now the students are achieving a higher level of math proficiency—faster, applying it in a work setting, and doing better as a result.

Standards don't matter for their own sake. They matter because they should help people meet requirements to get good jobs; to go on to higher education; to compete in the workplace; to succeed, prosper, and contribute. It shouldn't come as a surprise that students who understand the connection between their schoolwork and future earning power tend to do better in school. We can motivate students to learn about a subject by showing them how it pertains to the world of work—how they can use it to get a good job. And these standards should evolve and change as the requirements change.

There is no more important work than the development of math and science standards for our schools and children. To this end, I congratulate the National Council of Teachers of Mathematics. It has pioneered the development of education standards, making sure that they pertain to the real world, that they keep pace, and that students understand their importance.

We used to believe that not all students needed to achieve high levels of math. That may have been true when a good portion of our workforce and economy required relatively low-skilled manufacturing jobs. College-bound kids needed calculus but not everyone else. For everyone else, a calculator could take care of the rest—to balance a checkbook, to figure a tip, to compute a batting average. We talked ourselves out of math and convinced ourselves that it was hard and impractical. We tracked kids—those bound for college and those who didn't need math.

WE ALL NEED MATH

The bottom line is that today we all need math. Managers need math, technicians need math. Today's production workers need math. Math is a way of thinking, a thought process, a way of organizing the world and analyzing problems. Statistics is a basic part of any production process and any quality management process today. If you work on an assembly line at Ford or Chrysler, you're probably part of a quality circle or something similar, and you will have to use statistics to know how you are doing. You're running multimillion dollar robotic machines. You're getting paid more for your ability to manage that process than for manual labor. And the manual labor you might do will require a much higher degree of thinking and analytical skills.

Auto mechanics need math. Ten years ago, the normal training manual to be a certified mechanic was about an inch thick. Today, it is the equivalent of four or five major telephone books. You can't be a mechanic at a gas station today if you don't have the technical skills needed to work on today's cars. They may let you pump the gas, but there's not a big demand for that these days.

That's the case in every industry—fewer workers, higher skill levels, more responsibility, greater adaptability, increasing technicality. Those who have the skills will get to be a part of that workplace. Those who don't will be left behind.

No one should be left behind. No one should be denied a part of the workforce of tomorrow and the higher quality of life that will go with it. Which is exactly why *everyone* should be held to high, international standards. The fact that *all* children can learn, and will learn differently, make standards all the more critical for our disadvantaged children. I'm offended by the concept that children who are poor or minority should not be held to the same high standards. To the contrary, we should expect everything of our disadvantaged youth if we want them to be a part of this economy. Every now and then, I hear suggestions that standards will further disadvantage the urban poor. Standards aren't meant to be a system of reward, but a benchmark of what students need to know. These standards may appear higher for some, but they should be set, regardless, and our energies turned to helping *everyone* meet them. If the benchmark seems high for some, then let's be sure to give them the training they need to reach it. We won't be doing them any favors by simply lowering the bar and leaving them out of the workforce.

It makes sense that teachers help to lead this revolution. Standards give teachers the freedom to teach in different ways and to meet the individual needs of their students. When we know the goal, we can find different ways of achieving it. All children can learn, but they are all different and should be expected to learn in different ways. In turn, we should be able to adapt how we teach. This will strengthen the role of teachers as professionals who meet those needs and goals. Standards should not standardize curriculum—quite the opposite. They allow teachers to drive reforms in curriculum and teaching to meet the needs of their students.

Why should business and industry get involved in all this? For the simple reason that we can't afford not to—in the short term or in the long term. In the short term, we spend hundreds of millions of dollars annually on remedial education—teaching kids what they should already know. It's money coming out of the bottom line. In the long term, you won't have a workforce that can keep pace with your competition. You won't have the ability to react fast enough, to take advantage of the technology, to research your products. I'm not all doom and gloom—I'm optimistic because I know this will work—but if business does not get involved we will be a nation of hamburger-flippers.

HOW SHOULD BUSINESS GET INVOLVED?

How should business get involved? First, we should demand *standards*. We should demand better. If kids graduate from high school without meeting basic high school standards, we should send them back for more learning. Second, we should help educators, like the National Council of Teachers of Mathematics, understand what the workplace requirements are, as they have done in West Bend. We need to do more than complain: We need to help develop these standards in all key learning disciplines, hand in hand with educators. Third, we should help these standards evolve. In every business, the technology changes every year, as do the requirements. Standards should be an ongoing process of defining what it means to be the best. We should continuously help to "raise the bar." And finally, dedicate your resources to helping students reach that bar. Send a message to your community, educators, parents, and students, saying that just like a college, you have entry requirements. Help define what those are, and help people meet them.

The truth is that if we want to be competitive as individuals, as a community, as a company, or as a nation, then we, too, must make a commitment to education as a fundamental underpinning of our lives and our economy. Business has had a tendency to shy away from involvement in education, yet we have a direct interest in the ability of our schools to prepare students for the real world. The issue of standards in core subject areas should be an area where business and industry can contribute something specific that directly affects our ability to compete.

· · · · · · · · · · · · · · · ·

BIBLIOGRAPHY

Kearns, David T., and Denis Doyle. *Winning the Brain Race: A Bold Plan to Make Our Schools Competitive*. San Francisco: Institute for Contemporary Studies Press, 1988.

Kearns, David T., et al. *Xerox: Prophets in the Dark: How Xerox Reinvented Itself and Beat Back the Japanese*. New York: Harper Business Press, 1992.

The Public as Constituents

Rexford Brown

> *The public doesn't require any new ideas. The public is best
> served by the good, old-fashioned ideas it already has.*
> *(Henrik Ibsen)*
>
> *The public be damned. (William H. Vanderbilt)*

N̲O MAJOR school reform initiative can succeed without sustained public
engagement in questions like "What should our young people know and be
able to do in order to be healthy and productive adults?" Teachers and other
education professionals can come up with ingenious answers, but unless they
are debated and understood by parents and the larger public, educators' new
ideas will seldom take root.

Who is "the public?" It is everybody who isn't "us." Most of us want to
believe, especially when we have ideas about policy, that we are acting in the
"public interest." We spend a good deal of time talking about public opinion,
as if it were a single person with a single mind. The popular media poll the
public on its preferences and habits daily, reporting every nuance and fluctua-
tion. Depending on how one's efforts to change public institutions are faring,
the public seems either enlightened or benighted, gullible, or totally uninter-
ested in the truth.

Every group with a cause wants to influence public opinion, because the
public is also a majority or plurality of votes necessary for securing some policy
or program. Millions of person-hours are spent each week by nonprofit groups
plotting how to make this or that cause a household word, agonizing over how
to "get our message across."

*Rexford Brown is a senior fellow at the Education Commission of the States, currently
on leave to establish a K–12 charter school. He is the author of* Schools of Thought:
How the Politics of Literacy Shape Thinking in the Classroom *and numerous articles
about school reform.*

FOUR PUBLICS

The most sophisticated analysts of the public mind, who work in the private sector, understand that there is no single entity we can usefully call the public. They segment the population into dozens of minipublics, each with its own needs, buying habits, and agenda. Nonprofit organizations would be wise to do the same thing. Those of us interested in seeing mathematics more highly valued and more widely learned and practiced, for instance, might profitably ask ourselves who our minipublics really are and how we might engage them in dialogues about what matters most to us. From what I have observed over the last decade, the number one audience being addressed by the mathematics profession is the mathematics profession itself. Number two is policymakers, number three is business leaders, and number four is kids. Let me say a little about each, in reverse order.

Younger Children

Younger children are being addressed primarily through television specials, general children's programs, and regular mathematics-oriented programming on public television. By and large, this has been high-quality work that is much more interesting and sophisticated than mathematics usually is in school. I have been unable to find any data about which children, exactly, are watching these shows and whether (as I suspect) they are already interested in mathematics to begin with. But whether they are or not, the messages to younger kids are consistent: Mathematics is fun, interesting, and helpful; it is not "nerdy" to be interested in math; anyone, male or female, rich or poor, can learn math.

Older students get a different message. Few television shows about high school promote positive messages about mathematics. The most recent entry, *My So-Called Life*, portrays high school as at least irrelevant, if not inimical, to the lives of teenagers; and the only student remotely interested in mathematics is regarded by his peers as a nerd.

Business Leaders

Mathematics messages to business leaders have been aimed primarily at reinforcing the belief that U.S. productivity and competitiveness in the global economy depend on more and better mathematics instruction in school. International comparisons of student achievement in mathematics are often used to support this argument, since German and Japanese students appear to outperform American students, just as German and Japanese economies seem to outperform the American economy. This argument is probably most persuasive to people who think of economic strength in terms of numbers of engineers, inventors, and high-tech wizards in the labor force. It is not persuasive to people who view economic strength in terms of expanding markets, greater access to low-wage labor, and numbers of lawyers in the workforce.

Policymakers

Mathematics messages to policymakers also rely heavily on the argument that U.S. economic competitiveness is at risk. Over the last five or six years, I have heard it argued increasingly that mathematics education reform deserves policy attention because mathematics is a gatekeeper subject: Students who take algebra go on to success, while those who don't, don't. And those who don't tend to be poor and minority youngsters. Mathematics has thus been linked to the nation's equity agenda in a way that reading was linked to it in the 1960s and '70s.

The greatest boost to mathematics as a policy topic came about through the fact that the mathematics profession developed a complete set of standards just as state and federal policy leaders began to call for higher standards. Although most subject-area professional groups had been publishing increasingly more interesting and challenging statements about what young people should know and do, none had put anything together quite like the NCTM standards. These quickly became the model for disciplinary standards, now under development in science, geography, language arts, history, the arts and a number of other areas. When Governor Roy Romer of Colorado and other policy leaders stumped the country speaking about higher standards, it was the NCTM *Standards* documents that they waved and quoted.

The National Science Foundation's *State Systemic Change Initiatives* also kept mathematics visible to policymakers by providing large amounts of money to improve science and mathematics education. Many states have responded by, among other things, integrating the NCTM standards into their curriculum, assessment, and professional development structures.

Messages to policymakers about mathematics have been clear, constant for some years, reinforced by federal money, and supported by political muscle. To the extent that policymakers are a public themselves, represent the public, and influence public opinion, the mathematics community has done a good job of getting and keeping their attention.

Mathematics Teachers

In my experience, the public to whom most mathematics messages have been addressed is the education community itself, including mathematics teachers. If mathematics teachers do not know or understand mathematics as "the science of patterns and relationships" and so forth, and if they are unable to model best practice in their classrooms, all is for nought. And if some of the reasons why teachers do not know or practice what they should are related to the ways schools are structured and managed, teachers in other disciplines, principals, central administrators, and others throughout the enterprise need to know what they can do to restructure schools and districts so that mathematics can be taught in the best possible ways.

MUCH WORK REMAINS

My own observations of what's going on in schools indicate that much work remains to be done to get a coherent message about mathematics reform across to both mathematics teachers and the rest of the education community. A recent Public Agenda Foundation report lends support to my limited experience (Public Agenda Foundation 1993). Its poll of mathematics professionals and other educators found wide divergences of opinion about the scope of the problem, who can learn mathematics, and what should be done to improve the situation.

For instance, only 19 percent of math professionals rate the math proficiency of most students in their community as good or excellent; this is in dramatic contrast to 52 percent of administrators and 65 percent of principals in the survey. What's going on here? Are the principals and administrators out of touch, or are the math professionals exaggerating the problems in order to keep themselves in business? The school administrators also differ dramatically from college teachers about whether remedial mathematics is a pressing issue in college. The college teachers overwhelmingly say yes (69 percent), while only about a quarter of the principals agree with them.

On the question of who can learn mathematics, professionals also disagree among themselves, with a majority of school-level professionals being doubtful that all children can learn. Only 30 percent of secondary principals endorsed that view, for instance, compared to more than 80 percent of math supervisors and college teachers. Clearly, we have a schism between theoreticians and practitioners which, if not dealt with, will thwart mathematics reform. In my experience (Brown 1991), each side bases its opinion on entirely different data—the academicians on research in cognition and program results, the practitioners on what they see with their own eyes in their classrooms. To the practitioners, it is obvious that all students cannot learn mathematics, and research to the contrary is uncompelling. This schism is fundamental and profound and is not likely to be bridged with ordinary information, provided in ordinary ways.

Secondary principals are also a country mile away from math supervisors and college teachers on the question of whether students need to be taught the basic skills one at a time before they can solve complex mathematics problems, or whether they can learn to solve problems even before they know all the basic skills they might need. Almost four times as many supervisors and college teachers as principals believe in the latter approach. Considering that this is the approach supported by the new standards and that the principals' position represents common practice, much work remains to be done to get the principals on board. This schism, like the one between those who believe all students can learn mathematics and those who do not, is not likely to be bridged with a traditional information dissemination campaign. Something much more powerful is required.

THE PUBLIC AND MATH

The public is seldom asked for opinions about mathematics, specifically. Usually, it is polled about education in general or reform in general. In such polls, mathematics is included as one of the "basics." The Public Agenda Foundation report provides some indications of public opinion by including PTA members. They tend to line up with the principals. They do not think American students are doing badly, compared to students in other industrialized countries; only 41 percent support the practice of giving students problems before they have mastered the basics; only 41 percent say they believe all children have the ability to solve mathematics problems; and only 41 percent believe hands-on activities can improve math instruction. Unlike elementary and secondary principals, PTA members are not very aware of the NCTM *Standards*: Only 14 percent indicated they were "well aware." Only 10 percent of school board members indicated they were well aware of the *Standards*.

I do not want to make too much of a single poll, but these results square with my own experience. The general public may well say, in opinion polls, that it favors school reform and believes mathematics is important; but when it comes to the details of mathematics reform, it may well balk, along with many teachers, principals, and administrators. These groups hold decidedly less progressive views about mathematics reform than those of mathematics professionals in the academy and in national organizations.

ENGAGING THE CRITICAL MASS

The professional mathematics community has been successful in communicating with business leaders and policymakers through traditional comunication and dissemination vehicles. But it has not yet engaged a critical mass of the people who can make or break reform in any particular classroom, school, or school board meeting. Moreover, many of these people hold beliefs that cannot be turned around with traditional communications strategies. They must be engaged in long-term discussions about their beliefs and the ways in which institutional structures of schooling support those beliefs, long after science has shown they are mistaken.

One opportunity for deepening these discussions lies in the school restructuring movement that has been gathering steam alongside the mathematics education reform movement. Mathematics reformers must face up to the structural impediments that will limit their potential success. Structural reformers, for their part, are often too process oriented and unspecific with respect to the content of schooling. Neither can fully succeed without the other. It is time these two reform groups built a common argument for systemic change and communicated more in tandem.

Another opportunity for deepening understanding about mathematics education and why it must change awaits the necessary leadership. Sooner or later, this nation of many publics needs a large-scale conversation about the potential of all its citizens to learn whatever each is motivated to learn. Until that conversation, millions of children will grow up thinking they are not smart enough to learn mathematics or anything else of value. And we will remain a country unable to take advantage of the enormous wealth of human talent that lies all around us. Wouldn't it be gratifying to learn, twenty-five years from now, that by proving that all kids can indeed learn mathematics, the mathematics profession triggered that larger, and more consequential, national conversation?

· · · · · · · · · · · · · · · ·

BIBLIOGRAPHY

Brown, Rexford G. *Schools of Thought: How the Politics of Literacy Shape Thinking in the Classroom.* San Francisco: Jossey-Bass, 1991.

The Public Agenda Foundation. *Perspectives on Math Reform.* Pittsburgh, Pa.: WQED, 1993.

AFTERWORD . . .

The Workforce

Robert B. Reich

A FEW years ago, I wrote that Americans love to get worked up over American education because it is one of the few fields in which everyone can claim to have had some direct experience. The debates that have raged in the media, in Congress, and in school board and PTA meetings everywhere demonstrate that education reform is still a hotly debated topic.

American schools have traditionally mirrored the national economy. Our educational system at midcentury fit nicely into a high-volume, assembly-line economy. Now, a scant six years from the twenty-first century, our economy is changing dramatically but the form and function of the American education systems is not keeping up. The reality is that most schools have not changed for the worse; they simply have not yet changed for the better.

In this new boundaryless economy, what you earn depends on what you learn. And just as many of our best companies have responded to dynamism and change, so must our learning institutions. We can no longer think of education as something that happens to someone once. Learning is not an event; it is a process. A system where schooling, higher education, and training are segmented into specific time zones of our national life no longer meets the needs of the American economy or the American workforce. The heart of the Clinton administration's economic agenda is to revitalize, restructure, and integrate these distinct segments into a system of lifelong learning that constantly builds the skills and increases the knowledge of all our workers.

Dick Reilly and I have developed a running joke on our roles in the Clinton cabinet. He calls himself Secretary of Education and Labor, and I call myself

Robert Reich is secretary of labor. He was appointed to effect a revolution in the life-time training and education of our nation's workforce. He was on the faculty of the John F. Kennedy School of Government at Harvard University, served as an assistant to the solictor general, and headed the policy planning staff of the Federal Trade Commission. He has written extensively on the global economy and the U.S. workforce.

the Secretary of Labor and Education. You can no longer separate the two. You can't talk about future jobs if you don't get into the business of early education, curricula development, skill building—linking the world of learning with the world of work.

Over the last two years, Congress has passed a series of administration-sponsored programs that add up to a significant education reform package. They include an expansion of the Head Start preschool program; the Goals 2000 Educate America Act, which sets national achievement standards and boosts innovation; a school-to-work transition program; and a new program to subsidize college costs in exchange for community service.

WHAT'S THE DIFFERENCE?

Just recently, President Clinton signed the Elementary and Secondary Education Act, which targets more money to the nation's poorest children, encourages local educational reforms, brings more technology into the classroom, and puts more emphasis on preventing violence in our schools. Together, these programs add up to a solid record of achievement. But how do they differ from the plethora of reform schemes that have been circulating throughout the halls of academia and government ever since the Soviet Union launched the first sputnik in 1957?

The answer lies in the connection between the world of education and the world of work. Just as productivity can no longer be a matter of making more than what we already make at less cost per unit, productivity in education cannot be solely a function of the numbers of children who pass standardized examinations at a lower cost per unit. Therefore, these educational reforms encourage teachers and students to engage in the process of learning. Rather than prescribe exactly what should be learned and how and when the information should be doled out, we are about improving students' capacities to learn. Students and teachers are taking the initiative in deciding what is learned and when and how it is learned.

We are also establishing links between what is learned in the classroom and what is needed on the job. Goals 2000 encompasses a system for national, voluntary skill standards. Together with educational standards, that system is the cornerstone of our strategy to enhance workforce skills. A system of skill standards allows students to earn a credential that is portable and recognizable in the world of work. Job applicants will have a meaningful certification of their skill levels, which will improve their access to employment opportunities. Employers will have reliable, performance-based information with which to evaluate workers' abilities.

TRANSFORMING WORKPLACES

The School To Work Opportunities Act brings the workplace into the classroom and transforms workplaces into centers of learning. It encourages and assists the development of a nationwide school-to-work system that offers all students the opportunities and options they need to succeed in today's more demanding but more rewarding job market.

School-to-work programs create well-marked paths students can follow to move from school to good first jobs or to further education and training. These programs combine quality academic classes at school with hands-on learning and training at the work site. They are built on partnerships between schools and employers, and they emphasize local control to meet local needs.

Every school-to-work program contains three core elements: school-based learning, work-based learning, and connecting activities. School-based learning is classroom instruction based on high academic and occupational skill standards. Work-based learning is work experience, structured training, and mentoring at job sites. And connecting activities develop courses that integrate classrooms and on-the-job instruction, match students with participating employers, train job-site mentors, and build and maintain bridges between school and work.

At a minimum, students who complete a school-to-work program will receive a high school diploma or its equivalent and a skills certificate recognized by employers nationwide. Students whose programs call for additional schooling will receive a certificate or diploma recognizing the completion of one or two years of postsecondary education. Some students may enter a registered apprenticeship program or enroll in a degree-granting college or university at the conclusion of their school-to-work program.

These programs are based on existing models with a proven track record of success. The act offers states and localities limited federal funding—what might be described as venture capital—to refine and expand existing programs or start new ones. In July 1994, the first grants, totaling $43 million, were awarded to eight states to get their programs up and running. All states will be able to receive implementation grants by 1998.

COMMUNITY COLLEGES

Another resource with a proven track record is the too often overlooked system of 1500 community colleges across the country. They have long been the nation's unsung heroes of workforce investment—trailblazers on the pathways to lifelong learning.

Community colleges offer long-term skill development that goes beyond the hit-or-miss, one-shot efforts that characterize many traditional training programs. In addition, community colleges often team with local businesses to provide the type of learning that will be most useful on the job. Students who attend community colleges understandably want some assurance that the time and money they invest will actually lead to a job. Community-college links to local employers deliver on that promise.

Community colleges also serve men and women our current educational system frequently neglects: the older, so-called nontraditional student. In fact, 37 percent of students in community colleges are twenty-five years or older, proving once again the commitment these institutions have to continuous learning.

Nearly half of the part-time students in community colleges are also twenty-five years or older. Community colleges give us an existing infrastructure for a genuine system of lifelong learning that can upgrade a person's skills throughout his or her working life. And in the truest test of their value, community colleges are producing results. The median income of men with community college degrees is 26 percent higher than of those with only a high school diploma. For women with community college degrees, the statistics are even better. Their incomes are 33 percent higher than those of their peers who have a high school degree alone.

These statistics show the tangible benefits between what is learned in the classroom and what is needed on the job. That is the connection that we are trying to make with all our educational reform programs.

This country undoubtedly has the best system of higher education in the world. One-third of my students back at Harvard were from other nations. But for those 75 percent of our young people who do not graduate from a traditional four-year college, we provide one of the worst systems for getting from school to work. Our reform programs are aimed at changing that.

The challenge that remains for us on the eve of the twenty-first century is to shed the old, disconnected way of thinking about education and to welcome the change thrust upon us by new economic and social realities at home and abroad. That will not be easy. Change is unsettling.

But in the end, our options are limited. We can continue as we have, creating more jobs but not better jobs and reserving for the select few the benefits of our dynamic culture. Or we can commit ourselves to investing adequately in our own learning. In the world that awaits us and our children, developing these human resources is the only way for our economy to prosper and for all Americans to confront our common future with confidence.